The Craft & Art of Bamboo

Revised & Updated

30 Eco-Friendly Projects to
Make for Home & Garden

The Craft & Art of Bamboo

Revised & Updated

Carol Stangler

LARK BOOKS

A Division of Sterling Publishing Co., Inc.
New York / London

Editor: Janice Eaton Kilby
Art Director: Dana Margaret Irwin
Assistant Art Director: Hannes Charen
Assistant Editor: Heather Smith
Editorial Assistant: Rain Newcomb
Editorial Assistance: Anne Wolff Hollyfield
Photography: Sandra Stambaugh
Illustrations: Olivier Rollin
Development Editorial (revised edition):
Ray Hemachandra and Beth Sweet

Library of Congress Cataloging-in-Publication Data
Stangler, Carol, 1946-
  The craft & art of bamboo : 30 eco-friendly projects to make for home &
garden / Carol Stangler. -- Rev. and updated ed.
      p. cm.
  Includes index.
  ISBN 978-1-60059-339-0 (PB-trade pbk. : alk. paper)
  1. Bamboo work. 2. Bamboo construction. 3. Bamboo furniture. I. Title.
II. Title: Craft and art of bamboo.
  TT190.S73 2009
  745.51--dc22
                          2008018282

10 9 8 7 6 5 4 3 2 1

Published by Lark Books, a division of
Sterling Publishing Co., Inc.
387 Park Avenue South
New York, N.Y. 10016

Text © 2008 and 2001, Carol A. Stangler
Photography and Illustrations © 2008 and 2001, Lark Books unless otherwise specified

Distributed in Canada by Sterling Publishing,
c/o Canadian Manda Group, 165 Dufferin Street
Toronto, Ontario, Canada M6K 3H6

Distributed in the United Kingdom by GMC Distribution Services,
Castle Place, 166 High Street, Lewes, East Sussex, England BN7 1XU

Distributed in Australia by Capricorn Link (Australia) Pty Ltd.,
P.O. Box 704, Windsor, NSW 2756 Australia

The written instructions, photographs, designs, patterns, and projects in this volume are intended for the personal use of the reader and may be reproduced for that purpose only. Any other use, especially commercial use, is forbidden under law without written permission of the copyright holder.

Every effort has been made to ensure that all the information in this book is accurate. However, due to differing conditions, tools, and individual skills, the publisher cannot be responsible for any injuries, losses, and other damages that may result from the use of the information in this book.

If you have questions or comments about this book, please contact:
Lark Books
67 Broadway
Asheville, NC 28801
828-253-0467

Manufactured in China

ISBN 13: 978-1-60059-339-0

For information about custom editions, special sales, premium and corporate purchases, please contact Sterling Special Sales Department at 800-805-5489 or specialsales@sterlingpub.com.

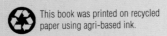 This book was printed on recycled paper using agri-based ink.

Dedication

To my father, Bernard Stangler

whose career as nurseryman and landscaper surrounded me with all things green and growing, and whose passion for fishing and the outdoors planted me firmly in the earth,

and to the memory of my mother, Inez Todnem Stangler

whose years as homemaker to our family of nine gifted me with the art of creating order and beauty, and whose lessons with needle and thread form the basis of my work.

CONTENTS

BAMBOO

offers us a balanced relationship with the land that meets the needs of the present without borrowing from the future or taking from the past.

In a world faced with global climate change and deforestation, the need for ecologically responsible materials is clear. Bamboo's rapid emergence in the West parallels the rise in global awareness of the planet's fragile ecosystem. In just a few years, researchers looking for alternatives have revealed bamboo's potential as a raw material in new eco-industries such as biofuels and bioplastics. New companies and businesses see bamboo's future broadening as an eco-responsible building material. The enormous potential of this magnificent plant is just starting to be explored.

Yet, already, the popularity of bamboo is amazing. Enter almost any store and you will see how entrepreneurs with their eyes on the future have worked this rapidly renewable material into innovative products and designs. Spun bamboo sheets and clothing, bamboo-veneered kitchen cabinetry, and

Bamboo's abundant evergreen foliage acts as a "carbon bank," absorbing about four times the amount of carbon dioxide absorbed by an equivalent stand of trees.

contemporary furnishings and accents are just a few items in a broadening range of affordable and high-quality bamboo products. In fact, next time you're in a new or renovated home or commercial building, look down at the floor—it could be bamboo!

Bamboo is one of nature's most productive renewable resources, with sustainable properties few other plants can match. Bamboo grows rapidly; it is harvestable in three to six years, compared to 40 or more years for trees. Bamboo captures and stores carbon, releasing 35 percent more oxygen into the environment than comparable stands of trees.

For the curious craftsperson, gardener, and environmentalist, sustainably harvested bamboo poles are now readily available for purchase, as are a wide variety of bamboo plants from nurseries. Whether purchasing poles or growing your own, *The Craft & Art of Bamboo* will give you the techniques and instructions needed to create 30 practical and beautiful bamboo projects for your home and garden. In the process of working with this "new" material, I trust you'll find bamboo as delightful and amazing as I have.

A Green Love Grows

How did I begin my own love affair with bamboo? I first stumbled upon bamboo groves in the early 1980s while harvesting kudzu vines to weave into sculptural baskets. I was awestruck by the giant, primitive stalks and dense growth. Then a friend took me to visit a 25-year-old bamboo grove along a slow-moving river in the southeastern United States. I was impressed by the grove's boundless energy—its growth along acres of river bank; the vibrant, tall culms bursting with leaves that caught the sunlight; and the almost impenetrable ground cover of accumulated dead poles and branches leaning against each other in all directions. Not knowing exactly what to do with bamboo, however, I put it in the back of my mind and waited for an opportunity to work with it.

In early 1990, I was asked to create an environmental sculpture for an Earth Day celebration in Atlanta, Georgia. I envisioned a piece called the Earthball that would measure 10 feet (2.9 m) in diameter and be constructed of natural, local materials woven together by Atlantans. The first challenge was figuring out what material to use for a spherical frame that would keep its shape while supporting hundreds of

Above: A forest of bamboo grows new culms and leaves every year, producing more biomass annually than a forest of trees.

Left: Beyond its spectacular beauty, bamboo is highly beneficial to humans and the environment.

Below left: Earth Day volunteers in Atlanta, Georgia, hold the bamboo framework of the 10-foot-diameter (2.9 m) Earthball.

Below: A nanako border lines a walkway in Japan.

9

pounds of woven vines. My mind's eye returned to the long, green culms of bamboo. Experimentation proved what I expected: Bamboo poles don't bend. Then I remembered a crude bamboo splitting device I'd seen workers using years before at the Bamboo Farm near Savannah, Georgia. We rigged up a similar device and split a truckload of culms into splints. Tying the splints together into long, thin bundles, we successfully wove and tied the bundles into a spherical framework. The process and the result were magical—a celebration of working together with one another and the earth's bounty.

Energized by the Earthball, I was ready to further experiment with bamboo. I received grant money to explore bamboo as a new craft material for the West, and soon after, in a moment of serendipitous timing, a Buddhist monk invited me to join him and a small group of Americans for three weeks' travel in Japan. Traveling across rural countryside and crowded cities, I saw bamboo—lots of it—made into everyday functional objects, such as brush brooms, supports for ancient tree limbs, and even long, bamboo tweezers, which a shopkeeper used to grab chunks of tofu.

I saw bamboo fences, too. I was enchanted by the simple beauty of their construction and amazed by the variety of applications used to build them and styles in which they were built. With the generous assistance of my Japanese guide and hosts, I returned to the United States with two all-purpose bamboo knives. I was ready to create with bamboo.

Months later, I held a reception to show the bamboo fences and screens that I'd made for an urban wildlife habitat. Guests came to celebrate bamboo. They drank kakicha tea from bamboo cups, ate nori rolls from bamboo trays, and viewed a craft material and art form new to many of them. Since then, I've continued to learn about bamboo and its remarkable eco-friendly properties.

Getting Started

You can choose from projects in *The Craft & Art of Bamboo* to make contemporary structures for your landscape, garden, and home. You'll find that what is strikingly beautiful is not always difficult to create! Most of the bamboo fences, screens, trellises, accessories, and other projects in this book are straightforward in construction once you've assembled the necessary tools and materials and learned a few basics.

To further inspire and energize your creativity, each chapter contains images of the work of contemporary bamboo artists. They are the pioneers in this new field who feel an affinity with bamboo and apply themselves to learning about it by trial and error, observation, and research. Their work gives us a glimpse of the diversity that arises when artists new to bamboo begin experimenting and using this rapidly renewable plant in their creative designs.

Try your own hand at crafting with bamboo. Relax in your favorite spot and look through the book, letting the projects and the beauty of this eco-friendly plant inspire you. Then visualize your landscape, garden, and home filled with your beautiful handmade efforts. Bamboo—sustainable, responsible, and green—may be the right material for you to redefine your own spaces and make a statement about protecting the planet.

Bamboo can be planted, harvested, and utilized locally, leaving little carbon footprint.

Chapter 2

Bamboo and Sustainability

East and West, North and South

BEYOND BAMBOO'S BEAUTY, THE MARVEL OF THIS MAGNIFICENT PLANT IS ITS ABUNDANCE AND VERSATILITY.

Perhaps no other plant has proved as useful to human beings as bamboo. Its earth-friendly properties and hundreds of uses have sustained and enriched the lives of people in many cultures around the world both throughout history and in modern times. For thousands of years, people have depended upon bamboo, and developed a sustainable, regional co-existence with the plant.

Bamboo in the East

Asian culture, past and present, is deeply intertwined with bamboo. Because bamboo's rapidly renewable properties have allowed it to flourish in so many locales for millennia, there's probably almost nothing in the East that has not been fashioned from bamboo. From arrows to dams, gates to hats, ladles to rakes, and umbrellas to zithers, bamboo has offered great utility and value in daily Asian life.

To the Chinese, bamboo takes on human characteristics. In literature and paintings, bamboo is the proverbial symbol for resilience in adversity: tough yet pliant, bending without breaking, a symbol of humankind's moral imperative to maintain strength and fortitude in times of turmoil and hardship. To the Japanese, human qualities represented by bamboo include enduring friendship, sincerity, and straightforwardness.

Top left: Outside this entrance in Japan, bamboo-faced planters hold New Year's greenery of pine boughs and bamboo.
PHOTO BY KONOMO UTSUMI

Top right: Bamboo scaffolding is being used during the construction of a Buddhist peace pagoda in Nepal.
PHOTO BY KONOMO UTSUMI

Bottom left: Bamboo railings guide worshippers at a temple in Japan.

Bottom right: Fishing boats of split and woven bamboo, sealed with a pine-like resin, rest on a beach in Danang, South Vietnam.
PHOTO BY MARLER SPENCE

Bamboo's Sustainable and Easy-to-Use Qualities

- Hundreds of varieties grow in widely diverse climates and ecosystems.

- Bamboo needs no seed collection or storage; it regenerates via underground rhizomes.

- It is harvested without the death of the grove and needs no replanting.

- Sprouts are a nutritious food source for humans and fodder for animals.

- Culms are lightweight, and the surface is waterproof.

- Poles split straight; flexible strips are easily made by splitting poles lengthwise.

- Poles may be joined with simple rope or strips of bamboo.

- A strong, sharp knife, used with skill, is all that's needed to work with bamboo.

Bamboo in the West

In the span of a few decades, bamboo has emerged to become an important plant for Westerners searching for green alternatives to nonrenewable resources. A high-yield grass whose entire biomass can be put to use, bamboo is being used in linens and clothing. Farmers are investigating growing bamboo for its nutritious sprouts. And green-thinking entrepreneurs are considering planting vast forests of bamboo to offset our carbon footprints.

The most successful bamboo product to date is in the building and construction industry:

Top: A variety of bamboo floorings are available to consumers who want to reduce their dependence on hardwoods.
PHOTO BY DYLAN MADDOX, COURTESY SMITH & FONG CO.

Bottom: Kitchen cabinets constructed with bamboo plywood offer a durable, handsome alternative to wood cabinets, yet are made from a rapidly renewable resource.
PHOTO BY MARK BOISCLAIR, COURTESY SMITH & FONG CO.

researched as a raw material for bioplastics and biofuel, in an effort to replace petroleum. Spun bamboo fiber already has entered the market and is being

Bamboo flooring is replacing oak, maple, and other hardwood floor materials. Bamboo flooring is manufactured by splitting and flattening mature bamboo culms into

River Cane Restoration

River cane is a North American bamboo essential to the preservation of Cherokee culture. Experienced artisans today continue the tradition of using cane to make blowguns and other crafts. However, the scarcity of river cane requires that craftspeople search far and wide for the raw material.

To ensure the continuation and reliable supply of this traditional resource, the Revitalization of Traditional Cherokee Artisan Resources (RTCAR) initiative was formed in 2004 thanks to a grant from the Cherokee Preservation Foundation to the Cherokee Studies Program at Western Carolina University in Cullowhee, North Carolina. One of its first projects identified and mapped mature stands of river cane surrounding the Qualla Boundary region of the Eastern Band of Cherokee Indians in western North Carolina. RTCAR is in the process of making arrangements with the landowners of those stands for selected Cherokee artisans to harvest river cane for their crafts.

Another RTCAR project is a research facility at Western Carolina University that conducts and supports scientific research on the ecology of river cane. Trial plantings are being studied to determine the best ways of propagating, growing, and harvesting this native resource.

Healthy river cane thrives under a sparse tree canopy next to a waterway.
PHOTO BY ADAM GRIFFITH

strips of equal dimensions. Processed and kiln dried, these flattened strips are pressed and glued together under high pressure to form planks. The planks are milled on the edges to form their tongue and groove slats. Finally, the surface is coated with a scratch- and abrasion-resistant finish. Bamboo plywood, which is used to make furniture, cabinetry, and other wood-intensive goods, is produced in a similar fashion.

Athletic flooring made of bamboo can now replace the traditional maple flooring of basketball courts.
PHOTO BY JOHN BENOIST, COURTESY SMITH & FONG CO.

Bamboo in North America

Only three known bamboo species are native to North America: river cane (*Arundinaria gigantea*), switch cane (*Arundinaria tecta*), and the recently discovered *Arundinaria appalachiania*. Of the three, river cane has flourished the most. For millennia, it was abundant in vast, dense forests called canebrakes along the edges of waterways in what is now the southeastern United States. Before 1500, river cane provided essential raw materials to native inhabitants, who depended upon it for crafting weaponry, housing, basketry, and other necessities for their survival.

Vast quantities of cane were required, however—60 to 100 mature culms were needed to craft a large storage basket—and the groves became depleted. So, Native Americans developed a practice of burning canebrakes on a rotation basis. When river cane near a village had been used for seven to 10 years, it would be burned over to allow its thick network of roots and rhizomes to re-sprout, regenerate, and fully mature.

Nonetheless, various factors have led to a drastic decline in river cane over the past few centuries. While river cane itself is not endangered, its habitat is. Although this bamboo species once was plentiful in North America, less than two percent of the original canebrakes remain.

Bamboo in South and Central America

Tropical guadua bamboo (*Guadua angustifolia*) is among the strongest and most durable of all bamboos. Prevalent in both South America and Central America, guadua is native to Colombia and Ecuador; it grows in giant forests in the Andes Mountains. Owing to its strength and great length—it grows up to 120 feet (36.6 m) tall—guadua is used in rural areas to make telephone poles, animal corrals, and irrigation pipes. It has long been a major

Cherokee artisans weave complex patterns into split river cane mats and baskets.

construction element in local housing for support poles and esterilla—bamboo poles that are split open, flattened, and used as lathing or siding. (See page 74 for the project "Side Table with Bamboo Esterilla.")

Deep in the jungles of Mexico's Yucatán Peninsula lies Yucatan Bamboo, the largest commercial bamboo plantation in the Western Hemisphere. Covering hundreds of acres, the plantation has 22,000 clumping iron bamboo (*Dendrocalamus strictus*) plants. The plants' mature poles possess a thick—almost solid—culm. These strong poles are used as is or milled into rectangular or square lengths, and they offer many applications as alternatives to wood, including as furniture, fences, and wall coverings. Yucatan Bamboo was founded in 1995 by eco-conscious agri-business entrepreneur Bob Gow, who wanted to produce an alternative to wood while employing local Mayans. Gow says the huge bamboo leaf canopy is a rich environmental asset that cleans the air by capturing vast quantities of carbon dioxide from the atmosphere.

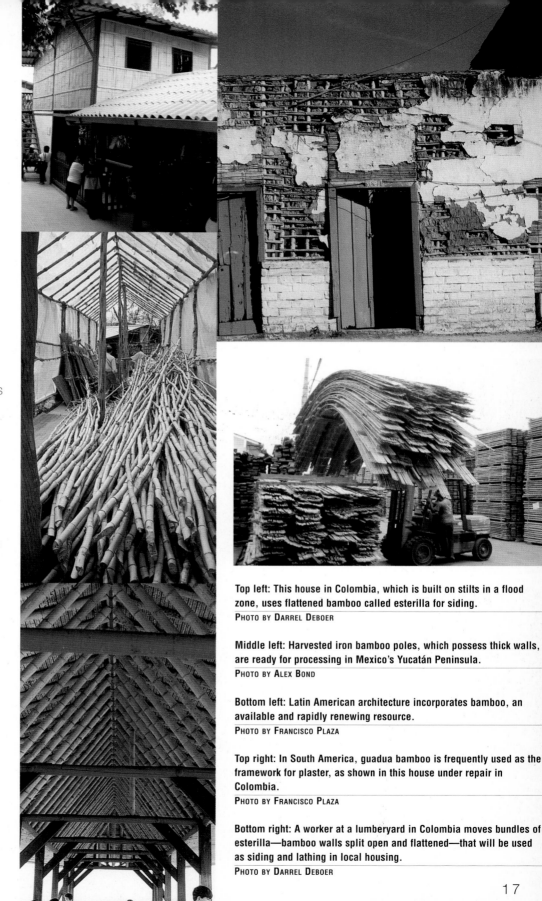

Top left: This house in Colombia, which is built on stilts in a flood zone, uses flattened bamboo called esterilla for siding.
PHOTO BY DARREL DEBOER

Middle left: Harvested iron bamboo poles, which possess thick walls, are ready for processing in Mexico's Yucatán Peninsula.
PHOTO BY ALEX BOND

Bottom left: Latin American architecture incorporates bamboo, an available and rapidly renewing resource.
PHOTO BY FRANCISCO PLAZA

Top right: In South America, guadua bamboo is frequently used as the framework for plaster, as shown in this house under repair in Colombia.
PHOTO BY FRANCISCO PLAZA

Bottom right: A worker at a lumberyard in Colombia moves bundles of esterilla—bamboo walls split open and flattened—that will be used as siding and lathing in local housing.
PHOTO BY DARREL DEBOER

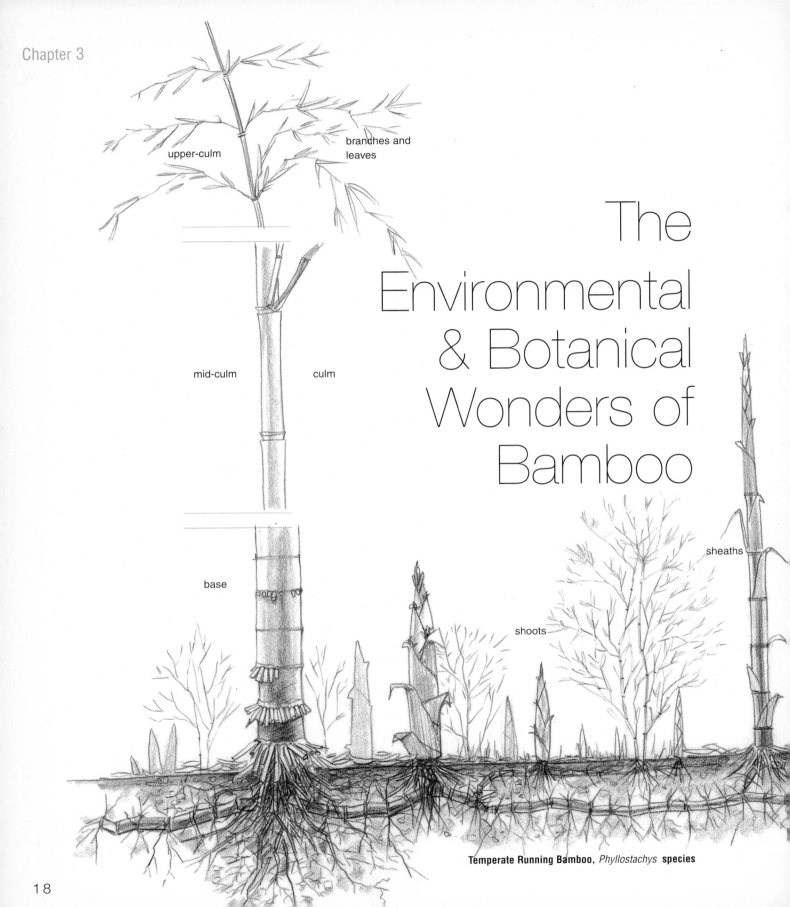

upper-culm

branches and
leaves

mid-culm

culm

base

shoots

sheaths

The Environmental & Botanical Wonders of Bamboo

Temperate Running Bamboo, *Phyllostachys* **species**

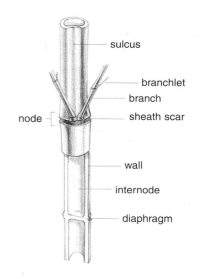

O NCE SEEN AS AN UNWELCOME INVASIVE, BAMBOO IS NOW HAILED FOR ITS ECO-FRIENDLY PROPERTIES. "RAPIDLY RENEWABLE" DESCRIBES ITS SPECTACULAR GROWTH RATE: BAMBOO REACHES 80 PERCENT OF ITS FULL HEIGHT, DIAMETER, AND LEAF CANOPY IN ONLY TWO MONTHS!

sulcus

branchlet

branch

node

sheath scar

wall

internode

diaphragm

shoots

2"

28"

rhizomes and roots

Bamboo flowers unpredictably, often inconspicuously, once every 30 to 120 years.

Bamboo is a giant perennial grass with a woody, jointed stem. More than 1,200 species of bamboo have been identified; they're classified according to the way their underground structures, called rhizomes, grow and spread. The rhizomes of tropical, or clumping, bamboo bunch together, multiplying outward into dense, circular stands. Temperate, or running, bamboo grows in cooler climates; its rhizomes grow outward in all directions.

How Temperate Running Bamboo Grows

The amazing growth spurt of the stem, or culm, is bamboo's most spectacular characteristic. Bamboo's dynamic growth is anchored and nourished by an underground network of rhizomes—the food-storage structure from which roots, buds, sprouts, and more rhizomes grow. Like the culm, rhizomes are segmented. They form a mat so dense that, even today, people seek refuge from earthquakes in centuries-old bamboo groves, knowing the rhizomes will hold the earth together. The rhizome network is shallow; 80 percent is in the top 12 inches (30.5 cm) of soil, and the remainder is in the next 12 inches (30.5 cm).

Fibrous roots grow from the nodes of rhizomes, seeking out and delivering water to rhizomes and culms. Pointed, tough buds grow from the rhizomes, pushing their way to the earth's surface to emerge as shoots. Husk-like overlapping sheaths surround each shoot, protecting the soft inner fiber. The shoots grow quickly and become culms, tapering upward to a thin tip. When the culm walls become strong enough to be self-supporting, the sheaths drop away.

Nodes, the raised joints that segment the culm into hollow compartments, emerge as the culm grows upward. Each node contains a diaphragm—a woody membrane that connects the walls and gives strength to the hollow cylinder. Internodes, the hollow segments of the culm between nodes, elongate upward gradually. Immediately below the nodes are sheath scars, which mark the previous attachment of sheaths.

Branches emerge from the nodes; branchlets emerge from the branches. Abundant bushy leaves grow from the branchlets. Generally, the upper surfaces of the long, narrow leaves are smooth, bright, and deep green in color, and the undersides are a light, matte green.

A thin wooden edging along a path will not deter bamboo from running. Here, rhizomes trying to grow new culms push up a brick.
Photo by Michael JamesLong

As the culm ages, its cellular tissue hardens. In three to five years, the culm is ready for harvesting and use. In an optimal environment, a grove matures in about 12 years, producing culms of maximum size.

Bamboo flowers unpredictably —and often inconspicuously— once every 30 to 120 years. Flowering saps energy but doesn't necessarily mean the death of the grove; the culms will likely grow back from their rhizomes or seeds. Bamboo seeds are roughly the size and appearance of grains of wheat. They germinate under the dead-leaf litter, and seedlings may eventually recarpet the grove.

This method of containing bamboo combines a plastic barrier with a rhizome trap. The trench will be backfilled with mulch or other light material.
PHOTO BY IAN CONNER

Containment and

Maintenance

Bamboo's most important ecological strength—its rapid regeneration—is challenging for people who have limited space—or neighbors! Left on its own, running bamboo runs. Only well-traveled roads and deep waterways—bamboo doesn't like "wet feet"—keep bamboo at bay. To be sure that bamboo is contained in your yard, install a rhizome trap or high-density plastic barrier—or both.

Each method involves digging a trench around the bamboo. Start by marking out the perimeter of the site, allowing about one foot of space

Avoid the difficult task of digging out rhizomes by installing a barrier when first planting bamboo.
PHOTO BY IAN CONNER

Green Contributions
Bamboo in the Garden

Bamboo is a strong candidate for your garden:

- It thrives without pesticides or herbicides.

- It uses little fertilizer.

- Planted only once, it multiplies on its own.

- Once established, it's drought resistant.

- It is tolerant of pollutants.

- It offers noise and dust abatement.

- It acts as a hedge, a privacy screen, or a wind break.

- It provides cooling shade and reduces heat buildup.

- Its poles can be crafted to use elsewhere in your garden and landscape.

Even a small grove of bamboo in the landscape offers a welcoming retreat.

between a fence or boundary line for future maintenance. A rhizome trap is a trench backfilled with light materials. The trap doesn't prevent the spread of rhizomes; rather, it facilitates cutting new growth. To make it, dig a trench 12 inches (30.5 cm) deep and wide, and fill it with mulch or sand. Each summer and fall, take a sharp, flat-blade shovel and thrust it into the trench, breaking up any spreading rhizomes.

The plastic barrier, a stiff polyethylene sheet about 1½ to 2½ inches (about 40 to 60 mm) thick and 30 inches (76 cm) tall, contains the rhizomes and acts as a root stop. Available from bamboo nurseries and suppliers, the barrier diverts the rhizomes over its top, where they can be clipped easily. To install a barrier, dig a trench 28 inches (71 cm) deep by one foot (30.5 cm) wide around the area. Place the barrier in the trench, with 2 inches (5 cm) raised above the ground. When using multiple barriers, overlap the ends and secure them together with a stainless-steel clamp; otherwise, the space between the barriers will be easily breached.

All bamboo, contained or not, needs regular maintenance. Remove the unwanted shoots in the spring; harvest mature culms annually; and remove dead and dying culms.

Historic Bamboos at *Biltmore*

by Bill Alexander, Landscape and Forest Historian, Biltmore Estate

As visitors drive along the approach road to the Biltmore House (www.biltmore.com) in Asheville, North Carolina, they are immediately immersed in a beautiful naturalistic landscape interspersed with extensive, shiny groves of bamboo. Now lush and green, it's hard to imagine that the estate's current 8,000 acres purchased by George Washington Vanderbilt beginning in 1888 were worn-out farms and deforested hillsides.

Historic Plantings

To transform the land, Vanderbilt invited Frederick Law Olmsted, a master of naturalistic design, to develop a restorative and sustainable land-use design. Olmsted undertook the challenge and implemented innovative landscape and forestry practices that continue to this day.

A unique element of Olmsted's restoration was his wide incorporation of exotic bamboos. Blending well with local plant species, the various bamboos offered tints and textures of foliage that created a constantly changing play of light and shadow, creating an aspect of mystery and bounty. Olmsted was so impressed with bamboos that his plantings at the Biltmore Estate are one of the earliest documented examples of the extensive use of hardy bamboos in the American landscape. Today, over 20 varieties grow in more than 40 locations on the estate.

Perhaps of equal historical significance was Olmsted's introduction of native river cane (*Arundinaria gigantea*) into a designed landscape. An entry in the 1891 landscape report stated that a thousand live canes had been collected from existing canebrakes and hauled to the approach road and other areas for replanting. In addition to its landscape effect, Olmsted also used the river cane to stabilize stream banks and pond slopes.

Frederick Law Olmsted, often referred to as the "Father of American Landscape Architecture," incorporated large quantities of bamboo at the Biltmore Estate more than a century ago. (Portrait by John Singer Sargent. Courtesy of the Biltmore Company.)

Biltmore Nursery

To supply the large quantity of bamboo and the millions of trees and plants necessary to reforest the degraded land, the Biltmore Nursery was established in 1889. In 1898 it entered the commercial market, selling a prolific variety of ornamental plants to all major centers in the United States. Among its offerings were 13 varieties of bamboo, described in the 1912 Biltmore Nursery Catalogue as "noble subjects" with "most enchanting results."

Chapter 4

Harvesting, Purchasing, and Preparing Bamboo

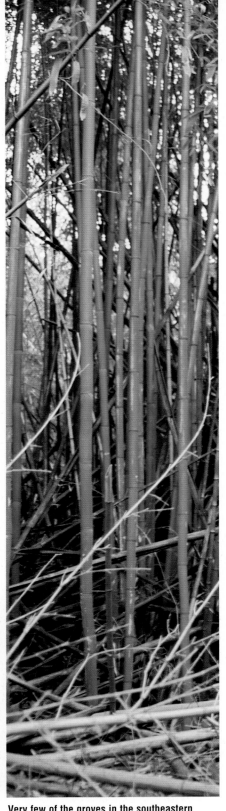

THE BEAUTY OF GATHERING YOUR OWN BAMBOO IS THE SATISFACTION GAINED FROM HANDLING A NATURAL MATERIAL AND CRAFTING IT FROM ITS RAW TO FINISHED STATE. OF COURSE, HARVESTING GIVES YOU A GOOD REASON TO EXPLORE A BAMBOO GROVE, WHICH IS A WONDERFUL ADVENTURE IN ITSELF. TIME REVERSES IN THE DARKENED INTERIOR OF THE GIANT GRASS FOREST, WHERE QUIET, SOLITUDE, AND THE GENTLE FLUTTERING OF LEAVES SURROUND THE HUMAN VISITOR.

Very few of the groves in the southeastern United States are properly maintained; the majority run rampant.

Harvesting

If you have access to a bamboo grove from which you can harvest your material, consider yourself lucky, but be prepared. Harvesting bamboo is for those who enjoy getting physical and interacting with nature in its unbridled state, and those who are willing to encounter anything in a grove from poison ivy to old refrigerators.

Getting Prepared

Before embarking on your expedition, be sure to obtain permission from the property owner. Next, assemble your harvesting tools (see the adjacent photograph), and wear long pants, a long-sleeved shirt, gloves, sturdy shoes, and eye protection. Don't underestimate the value of covering your arms. One hot spring day, I wore a T-shirt to harvest and haul bamboo, then spent the night nursing a fiery red rash on my forearms. Tiny silica crystals on the surface of new growth culms had pierced my skin like fiberglass. Ouch!

When to Harvest

Winter and very early spring are the natural times of year to harvest temperate bamboo. Growth is complete for the year, and roots and culms are in a resting cycle. Your work will be easier, too. Vines and overgrowth amidst the culms have

Harvesting tools include a collapsible tree pruning saw, machete or loppers, and hand pruners. Bring rope, bungee cords, collapsible sawhorses, and a tape measure, too.

lost their leaves, snakes and insects are dormant, and the air is cool and brisk.

Selecting Older Culms

The underlying principle of harvesting bamboo from a grove is the same as maintaining a healthy garden: prune out old growth to make way for the new. Generally, the culms of temperate-climate, running bamboo grow old and die after five to eight years. It's best to harvest older, mature culms because their cellulose walls have developed strength. Also,

Older culms may be yellowed or scarred.

Another technique is to look upward in the grove canopy for dying, leafless tops. They're not always easy to see, however, and it can be difficult to distinguish which top is connected to which culm. In this situation, it's very helpful to work with a friend. While you're inside the grove, have your spotter positioned outside to direct you to areas with dying tops. Then shake the culms until your friend confirms that you're shaking the one with the leafless top. You can also harvest standing, dead culms. Just check to make sure they're not cracked, split, or rotten.

Cutting and Hauling

Once you've located an appropriate culm to harvest, check to make sure that you have a viable "escape route," a straight path out of the grove starting from the base of the culm. Take into account that both you and the 12- to 40-foot-long (3.6 to 12 m) culm may have to squeeze and maneuver through a thick bamboo forest.

much of the sugar in the walls has converted to starch, making them less attractive to boring insects. Older culms also contain less moisture, making them lighter to transport.

How do you distinguish old culms from new in a grove? Look closely. If a culm has a bright green, relatively clean surface and dried sheaths encircling its base as shown in the above photograph, it's probably a first- or second-year culm. If the culm is dark or dull green, perhaps with brown, yellow, or tan areas, chances are it's older and ready for harvest. Quantities of mildew and dirt on the culm are also a clue to older age.

After selecting the culm and confirming that you have a way out, crouch down and clear away the *duff*, the dried and decaying bits of bamboo leaf, sheath, and stalk, from the base of the culm. Use the coarse-tooth pruning saw to cut the culm as close and even with the ground as possible. This practice eliminates stubs that could trip you, stick you, or hold water that breeds mosquitoes. If the base is 3 inches (7.6 cm) or greater in diameter, first make a back cut about one-third of the way through to get a clean cut and to keep the saw from getting stuck.

After cutting, fold up the saw and secure it in your pocket. Haul the culm, base first, through the escape route. Haul the culm into the open where you've positioned two sawhorses. Remove branches with a machete or loppers. If you've already planned your bamboo project and its components, you can measure and cut the lengths you need. Otherwise, a good rule of thumb is to cut the longest length your vehicle and storage area will accommodate.

Grove Etiquette.

More often than not, the branching top third of the culms will not be used in your projects. What to do with all that brush? Since branches and culms break down slowly, leaving brush to decay haphazardly in the grove is an unhealthy environmental practice. The debris restricts the air movement necessary to ward off fungus and insect attacks. It also hampers other visitors' efforts to harvest and tend the grove. For long-term bamboo brush com-

posting, it's best to designate an area adjacent to the grove for the orderly stacking of brush and culm debris.

Transporting

Moving bamboo to your work space can be tricky. Unlike milled lumber, bamboo slides and rolls, making it a hazard on the highway if it's not snugly tied to your car. Once I was in a hurry to drive a short distance from grove to workshop, and I did a sloppy job of lashing 12-

foot (3.6 m) lengths of bamboo onto my car's roof rack. As I drove downhill, the poles slid off, and I watched with horror as they went flying like torpedoes onto the street ahead of me. Lesson learned. Equally surprising is the weight of green bamboo, so be prepared for that too. On one expedition, the roof of my tiny station wagon caved in under the weight of my harvest, and I agonized every time my car roof thumped in and out for the 75 miles (120 km) home.

Bundling is an effective remedy for these challenges. As shown in the photographs to the left, use a rope or several bungee cords to tightly wrap together a manageable number of poles. Create bundles that you can lift and carry by yourself or with a friend's help. Tightly secure the bundles to your vehicle and to each other. Use lots of rope and double-check that the load won't shift.

Purchasing

If you don't have access to a grove, you can buy bamboo poles from a number of sources. For small projects, you can often find 4- to 5-foot (1.2 to 1.5 m) bamboo stakes, ½ inch (1.3 cm) in diameter, sold singly at import stores, garden centers, and home improve-

ment stores. For larger projects, there are many sources for poles. See page 160 for supplier information.

Western companies import a variety of sizes of bamboo poles, primarily *tonkin* (see page 33), for the nursery and decorative trades. Imported from China and southeast Asia, the poles have been sorted, sized, bundled, and fumigated before they reach Western warehouses. Poles 1 inch (2.5 cm) or less in diameter are sold in bundles of 50 or more, while poles 1 to 3 inches (2.5 to 7.6 cm) in diameter are sold in bundles of 5 to 25. The largest poles, 4 to 6 inches (10.2 to 15.2 cm) in diameter, are often sold singly. Many suppliers offer splitting and cutting services.

As non-native species of bamboo have established themselves in Western groves, a variety of mature culms are now being harvested annually, primarily running bamboo in the United States and Europe, and clumping bamboo in Australia. Poles grown, harvested, and shipped within domestic borders are not required to be fumigated. Some growers cut to order and ship green.

Although not made of bamboo, *reed fencing* is often carried by bamboo importers. Stalks of heavy grass are bound together with wire and sold in 8-foot (2.4 m) rolls in widths of 4 and 6 feet (1.2 to 1.8 m).

If you buy by mail or on the Internet, be sure to factor in shipping costs when placing your order. In the United States, parcel delivery services will deliver bundles up to 8 feet (2.4 m) long. For shipments exceeding these dimensions, shipping is by common carrier, i.e., trucking companies.

Storing

Bamboo should be stored out of the weather in a covered area such as a garage, shed, or dry basement. Use pallets or sawhorses to keep the poles off the ground to prevent moisture from wicking up and causing rot. I have used a variety of found materials to create a wall of bins, baskets, and containers to organize and store my bamboo. Keep in mind that the greater the variety of bamboo you accumulate, the more choices you have during the creative process. Collect, collect, collect! Save your scraps. Keep your bamboo stored where you can admire it. Let it inspire you.

Drying and Weathering

As freshly harvested bamboo dries, the chlorophyll fades and the uniform forest green may become shades of green and yellow with streaks of brown. Scarring, caused by culms falling across one another, or raised dark spots may be noticeable. These harmless sprinkles, known as "sesame," are caused by organisms that alter the surface of the outer wall.

After about six months of outdoor exposure, bamboo dries completely, turning a uniform light tan. As bamboo structures are exposed to the elements, they naturally become weathered. Sun and rain break down the outer layer of silicone, and the surface becomes pitted. Mildew settles in and dirt accumulates.

Annual cleaning and sealing greatly increases the life and appearance of bamboo structures; see chapter 5 for suggested annual maintenance. If, however, you choose to allow an outdoor bamboo structure to age naturally, it will no doubt become one with its surroundings, returning gracefully to the earth.

Bamboo is best stored in a covered area.

As fresh bamboo cures, its color changes from green, to brown, to tan.

Insect Damage and Control

Where there is food, there are bugs! The sugar in the walls of freshly cut bamboo may become a meal ticket for wood-boring insects. The mature

Over time, the sun may dry and bleach bamboo to a silver gray. In humid climates, moss and lichen may take hold.

insects deposit their eggs in the harvested culm. As the eggs hatch, the larvae chew through the inner walls and finally emerge through the outer wall.

Fumigation, best done in industrial settings, is the only sure way to prevent and eliminate bugs. The risk of infestation can be lessened however, if you harvest in late winter, harvest mature or dying culms, and store harvested poles upright with their base ends down until

Pinprick-size holes in the bamboo and tiny mounds of finely ground powder are clues to insect infestation.

drying is complete. All three practices help keep the sugar in the poles at their lowest level while curing. Sweating the poles will also reduce the risk of insect damage; see chapter 5 for this technique.

If your stored bamboo becomes infested, discard the worst. Poles or other pieces that are slightly infested can be used for supports, markers, and simple temporary structures lasting only one or two seasons. If borers are present in poles or projects you've already completed, you can kill the larvae with chemicals or by freezing them. Put short lengths in plastic bags and place them in the freezer overnight. This kills the chewing larvae but not the eggs, so you'll need to repeat the process several weeks later after any eggs have hatched. For long poles or large, finished projects, chemical insecticides for boring insects are effective and widely available in garden stores. Follow the package directions carefully to minimize your exposure to the chemical and its impact on the environment. Work outdoors, and wear protective gloves and a respirator. Brush the chemical solution onto the outside walls of the infected lengths. If the bamboo is split, apply it to the pulpy inside walls. Let dry overnight, then apply sealer.

Cleaning and Fungus Removal

Whether bamboo is harvested or purchased, washing with soap and water removes dirt and mildew from bamboo and brings out the luster of its shiny surface. To remove the worst of the grime, line up poles against a wall or across two sawhorses, and use a hose with a high-pressure nozzle to blast off the dirt. Spray with a nonsudsing, mildew-removing cleaner, then follow with a water rinse and let dry.

When bamboo is surrounded by stagnant air and dampness, patches of mold and fungus may grow on its surface and on the inside walls of split bamboo. To discourage such growth, regularly rotate your materials. In closed spaces, use a dehumidifier to collect moisture and a fan to keep the air

For a more thorough cleaning, especially around nodes where dirt collects, use a scrub brush with the cleaner.
PHOTO BY SANDRA STAMBAUGH

moving. You can spot-clean mold and fungus by moistening a cloth with isopropyl alcohol and rubbing it over the affected area. Alcohol kills the organisms, disinfects the area, and evaporates quickly.

Bamboo Poles Available for Purchase

COMMON NAME/ GENUS & SPECIES	Yellow Grove *Phyllostachys aureosulcata*	Vivax *Phyllostachys Vivax*	Red Margin *Phyllostachys rubromarginata*	Black *Phyllostachys nigra*	Robert Young *Phyllostachys viridis 'Robert Young'*
TYPE	temperate, running	temperate, running	temperate, running	temperate, running	temperate, running
CULM DESCRIPTION	culm is vibrant green with yellow sulcus; some culms zig-zag at the base	unpronounced nodes; resembles Giant Bamboo but internodes are slightly off-center and somewhat curved; old culms mature to light gray-green	smooth, medium to dark green, long internodes of 16 inches (40.6 cm) or more; in larger culms, walls bulge asymmetrically above node	slender, chocolate brown to ebony black; green the first year, then turns dark	initially sulphurous green, turns gold-yellow by end of year; random green stripes along intern-odes
WOOD QUALITY	not of superior quality, but good for general purposes	very good quality	very good quality	thin-walled but durable; used to make furniture and decorative objects	very good
ORIGIN	China	China	China	China; also intro-duced to Japan long ago	China
DIAMETERS AVAILABLE	1 to 2 inches (2.5 to 5.1 cm)	2 to 3 inches (5.1 to 7.6 cm)	1 to 2¼ inches (2.5 to 5.7 cm)	½ to 2½ inches (1.3 to 6.4 cm)	2 to 3 inches (5.1 to 7.6 cm)

Iron **Dendrocalamus strictus**	Tropical Black **Gigantochola atroviolacea**	Tonkin **Pseudosasa amabilis**	Henon **Phyllostachys nigra, 'henon'**	Moso **Phyllostachys pubescens**	Giant Timber **Phyllostachys bambusoides**
tropical, clumping	tropical, clumping	tropical, clumping	temperate, running	temperate, running	temperate, running
slender, thick to almost-solid wall	black with occasional faint green strips that turn tan when dry	straight with smooth nodes, cures to blond or tan	mature culm covered with green/gray waxy coating; may be removed to reveal a rich hue	can grow large, up to 7 inches (17.8 cm) diameter	straight, even, thick-walled
hard, dense, important for construction	sturdy, used for fine furniture and musical instruments	thin-walled but tough, exceptionally resilent and strong	thin-walled but strong; almost as good as giant timber bamboo	wood relatively soft, but poles are very versatile	hard, versatile, used in construction and crafts
India	Java, Sumatra	China	China; also introduced in Japan long ago	China; also introduced to Japan long ago	China, India; also introduced to Japan long ago
1 to 1½ inches (2.5 to 3.8 cm)	2 to 4 inches (5.1 to 10.2 cm)	¼ to 2 inches (6.4 mm to 5 cm)	2 to 3 inches (5.1 to 7.6 cm)	2 to 7 inches (5.1 to 17.8 cm)	1 to 5 inches (2.5 to 12.7 cm)

33

Chapter 5

Tools, Materials, and Techniques for Working with Bamboo

T O WESTERNERS ACCUSTOMED TO WORKING WITH UNIFORM LENGTHS OF MILLED LUMBER, BAMBOO POSES NEW CHALLENGES. THE BEST APPROACH IS TO WORK WITH THE NATURE OF THE BAMBOO, HIGHLIGHTING ITS FORM AND TEXTURE, AND ENJOYING ITS IRREGULARITIES AND QUIRKS.

Planning and Design

Start by choosing a project that interests you, suits your landscape, garden, or patio, and fits your available time. Building with bamboo can be surprisingly labor intensive, and large projects should be well planned. For any project, calculate the number of poles needed and account for their diameter and length just as you would with lumber. Figure in extra poles to replace the unsuitable ones which may be crooked, split, or otherwise unusable. When making trellises or other structures, lay out your design on a large flat surface such a patio, and adjust and replace poles to get the best design and fit before you cut.

Cracking

Know that bamboo culms can sometimes split when their walls dry and the air trapped inside expands with changes in temperature and humidity. I cringe every time I hear a "BOO!" coming out of my studio. Accept it: cracking happens. There is a technique, however, that may lessen the chance of splitting. Before assembling a project, use rebar or threaded pipe to knock out node membranes inside the culm. This gives heated air inside the chambers a place to escape.

Top to bottom: Japanese saw, plastic pipe saw, hacksaw.

Cutting Bamboo

Bamboo has a smooth, hard outer surface. Tiny silica crystals form a glasslike coating which repels rain but is difficult to penetrate. Beneath this outer wall is an inner wall of tough cellulose fibers. Your equipment and cutting technique will need to account for these factors.

SAWS AND SAWING TECHNIQUES

To cut directly across a culm, you'll need a sharp saw. Many types of saws will work, but for rough cuts I rely primarily on a hand saw designed for pruning trees. Fine cuts are best made using a pull saw with a thin blade and 22 teeth per inch (2.5 cm). The cutting teeth must saw into the fibers cleanly, so as to not shred the culm's sur-face. It's essential that the saw teeth be sharp and complete, no easy task since bamboo's hard silicone surface quickly dulls blades and damages teeth. Western hacksaws with metal-cutting blades work well, as do Japanese saws, which are more heavily weighted with blades specifically designed to cut bamboo. Power saws also work well, and the fine teeth of a ceramic disc blade give a good cut.

It's much easier to saw cylindrical bamboo when you use quick-release clamps. Use the clamps to secure bamboo to your workbench, and making precise cuts will become a no-struggle, no-juggle situation. Another approach is to use a Japanese cutting box. Designed for use while sitting

Photo 1: When using a Japanese cutting box, rotate the pole to get a clean cut.

Photo 2: To eliminate "feathering," hold the end of the bamboo while you cut.

Photo 3: Using a miter box to cut bamboo at a 45- or 90-degree angle.

on the floor, the box can also be placed on a bench to achieve a comfortable working height. The v-shape of the box provides a secure groove for the culm as shown in photo 1. When not in use, the box holds tools.

If your saw is too coarse or dull, the outer wall of bamboo you cut may shred or "feather." As the saw finishes the cut, the culm may break too soon, leaving a ragged edge on one side of the cut and the inner wall exposed on the other. For initial

Western-style "push" saws such as hacksaws are primarily "push to cut," meaning the teeth cut on the stroke moving away from your body.

Japanese bamboo saws are "pull to cut," cutting on the stroke moving toward your body.

harvesting or rough cuts, feathering may not matter. For fine cuts, however, you'll want to achieve a clean, sharp edge. If you're using clamps, saw about three-quarters through the culm. Next, use your other hand or a support to carefully hold the extended end (see photo 2). Continue to simultaneously saw, while supporting the extended end, until the blade cleanly penetrates through the outer layer.

If you're using a cutting box, first cut through the hard outer layer and the wall. Then, as you saw, rotate the bamboo so the saw is always cutting into the fibers. This helps you prevent tearing the outer fibers. Keep the blade straight so that when you complete the rotation the beginning and end of the cut will meet.

You may wish to cut 45° angled ends, which give an open, elegant feeling to a project. Also, if

you wish to butt two pieces of bamboo against each other at a 90° angle, it's important that your cuts be correct. To avoid problems, use a simple miter box, which is an open-ended "jig" marked with slots set at 45° and 90° angles to guide your saw. Mark the desired angle on the bamboo, and place it in the miter box. As shown in photo 3, line up the mark with the appropriate set of slots, clamp or hold the bamboo securely, and saw. For projects requiring less uniformity or precision, diagonal cuts can be made without a miter box, but it always helps to mark the diagonal angle before you cut.

USING A JAPANESE KNIFE

A Japanese knife is an excellent tool to "manicure" bamboo when you need to make fine adjustments in circular or square openings in order to fit poles into those joints.

Japanese knives and the technique for using them is quite different from the Western-style pocket knife. The blade is just 1 inch (2.5 cm) or so long and is placed on the diagonal at the end of a steel handle.

To cut with the knife, hold it near the blade in your dominant hand. Grip the bamboo with your other hand, near the hole you wish to scrape. If the pole is long enough, tuck it under your arm against your side to stabilize it. Now use the thumb of your non-dominant hand to push the knife

Japanese knives for splitting and cutting.
KNIFE SHEATH ON RIGHT BY TAKEI AKINORI

Photo 4

down into the fibers. Unlike Western knife techniques, the hand holding the knife applies no forward pressure. Instead, the thumb pushes downward, lending more control of the very sharp blade (see photo 4). When you've cut around one-quarter of the curve, you'll notice that the blade will not properly cut through the next quarter curve. To continue carving, you'll need to readjust the position of the pole, your hands, or the knife so the blade cuts through the fibers.

Splitting Bamboo

Artistic possibilities expand when you work with split bamboo. The halved side can be placed against flat surfaces, and shorter split lengths make good wall coverings and inlay for bamboo "mosaics."

Freshly harvested green bamboo is easier to split than dried and cured bamboo. Green bamboo shrinks, however, and

Split bamboo is woven between bamboo stringers. Design by The Bamboo Fencer.
PHOTO BY DAVE FLANAGAN

Detail of a split and woven bamboo fence panel.

DESIGN AND PHOTO BY THE BAMBOO-SMITHS

The front of this shed is faced with split bamboo and tree bark.

DESIGN BY MICHEL SPAAN

split green bamboo shrinks even faster. It's easiest to make pegs, handles, and other fittings from green bamboo, but allow them to dry thoroughly before incorporating them into a project.

SPLITTING TOOLS AND TECHNIQUES

It's easy to split bamboo by using a splitting knife and a rubber mallet. A good splitting knife has a strong blade and a sharp double bevel. Splitting knives vary in size and heft; use a blade 8 inches (20.3 cm) long to split large-diameter and woody lengths, and a shorter, thinner blade for finer work. An all-purpose splitting knife is the Japanese *take ware*, with a blade 8 inches (20.3 cm) long, hand-forged of high-carbon

steel. Any strong, double-beveled knife works well as long as it's kept sharp. You'll also need a rubber mallet to drive the knife through the culm. Don't be tempted to substitute a steel-headed hammer; it will damage the top of the splitting blade.

Successful splitting requires two things: good initial placement of the blade and keeping the blade along the same vertical grain throughout the length of the culm. First, cut the pole to the length desired. Next, determine which end you will start your split. The accepted school of thought is to split from the top of the culm toward the base, "top to bottom always." On mid-culm lengths, however, it can be difficult to

determine which end is which; in that case, look and feel for the sheath scar of the node. It's always toward the base.

To determine the placement of the split, put one end on the floor or a low surface and eyeball the length from top to bottom. Rotate the pole; it may be very slightly curved. The idea is to split along the axis of the pole, which will yield two somewhat straight lengths, as

A worker at the Bamboo Farm, Savannah, Georgia, splits a pole through a blade welded onto an angle iron.

opposed to two curved lengths. The split halves may arch slightly but will flatten when attached to a hard surface with lashing, screws, or nails. At the top end of the pole, mark the point where you plan to begin the split.

To split, stand the pole on end on a chopping block or brace it in a corner. Place the knife blade over the starting mark you made, with the end of the blade protruding several inches beyond the pole. Using the rubber mallet, knock the blade securely into the culm (see photos on right). The force of the strike and the thickness of the walls will determine the depth of the split. If the pole is thick-walled, the blade may drive into the grain only ¼ inch (6 mm) or so. If the culm is thin-walled, the blade may go deeper.

Using the mallet, hammer the blade through the full length of the culm, keeping your eye on the split to keep the two sides equal. If they start to split unevenly, adjust the angle of the blade by gripping the handle and twisting the sharp edge of the blade toward the side that needs to be thinner and away from the side that needs to be thicker. Maintain this angle as you hammer the blade through several nodes or until the sides become equal again. Then straighten the blade and continue splitting. Don't expect

Clockwise from top: machete, splitting knife, *take ware*, rubber mallet, four-way splitter.

perfection every time! Some poles are tough or seem to have minds of their own. Practice is always the best teacher.

Some artisans skip the mallet and use only a knife and a pounding or "tapping" action. Position the blade across the end, and with one sure and swift motion of the wrist, lift the

blade and drive it securely into the culm. With the blade now fixed in the bamboo, tap the culm's base end on a hard surface. At each impact, the blade will cut through the culm. You can vary the stroke from a light tap to a hard pound. Angle the blade to achieve even halves.

Making Splints and Removing Pith

Splints are narrow lengths of bamboo used to make pegs, chopsticks, woven fence panels, mats, and baskets. To make splints of any width, use a splitting knife and mallet to split a length of bamboo in half. Remove the diaphragms with a hammer or knife. Split the half-round length, and split each half again. Repeat until the splints are the desired width.

You can use a hammer to knock out diaphragms.

To make splints of ¾ inch (1.9 cm) wide or less, you'll need to have more control over the splint. The Japanese technique of pinching the splint while twisting the knife can help. Near the top of the splint, hold the edges between your thumb and index fingers. If the splint is long enough, tuck the other end under your arm against your chest to steady it. Hold the splitting knife in your other hand and rock it back and forth on the end where you want the split, pushing the blade in so it

When splitting or removing the pith to make splints, wear leather work gloves until you've mastered the techniques. A blade that's sharp enough to cut bamboo can easily cut your fingers.

separates the fibers. Lift it out, squeeze the splint between your thumb and index finger, and with a short, quick flick of the wrist, embed the blade further into the fibers. Release the "pinch" and twist the knife blade. This separates the fibers and allows you to move the knife further down the splint. Continue to pinch, strike, and embed the blade, and twist, keeping the split along the same grain.

To achieve a finer, more flexible splint, you can remove a third or more of the pithy inner wall. Holding one end of the splint,

Various sized bamboo splitters hang in a basket-making studio in Kyoto, Japan.

place a sharp splitting blade lengthwise across the other top end. Rock it back and forth to embed the blade into the pulp, then twist it, breaking apart the sheet of pulp along the grain. Continue to top the blade down the splint and twist. If the splint is still not fine or flexible enough, remove another sheet of pulp. Use a knife or shaver to smooth the inner wall and edges.

You can also use a commercial splitter; it's made of four or

A four-way splitter makes splitting bamboo easy.

more razor-sharp, hardened steel blades affixed in a cone shape that's attached to two handles. The choice of the number of blades depends upon the diameter of the bamboo and the desired width of the splints. To use the splitter, center it at the thinnest end of the culm. Tap it or use a rubber mallet to drive the blades into the culm. Firmly grip the handles and tap the bottom end of the culm against a hard wooden surface, pushing the splitter down through the nodes.

Materials and tools for sanding and smoothing. Left to right: sanding block, sandpaper, steel wool, flat rasp, round rasp, shaver.

I bought a power drill and drywall screws. Although blatantly "un-Japanese," screws work quickly and efficiently to join bamboo to bamboo and bamboo to lumber. Screw heads can be covered by lengths of split bamboo or decorative ties. The best type of screws to use with bamboo are thin-diameter, with threads close to the head, a thin shank, and a Phillips head. Drywall screws, which range from 1 to 3 inches (2.5 to 7.6 cm) long, work well. For outdoor work, use galvanized decking screws. Their shafts are narrow and range in length from 1½ to 3½ inches (3.8 to 8.9 cm). Wood screws aren't recommended, because their shank broadens toward the head; this causes the bamboo to split if the pilot hole isn't as wide as the shank. If you want a screw head to be flush with a rounded culm, use a countersink drill bit to hollow out a shallow crater before you drive in the screw.

Smoothing

The edges of split bamboo can be surprisingly sharp, and the pain of a bamboo sliver under a fingernail isn't soon forgotten, so wear work gloves. For finer projects in which you handle the object, use a beveled knife or shaver to smooth the sharp edges of the inner wall. A few quick strokes, and splintery edges are gone! Additional smoothing with a sanding sponge makes the bamboo feel like silk. Use a curved or round rasp to smooth the edges of holes drilled for pegs or poles.

Joining and Attaching with Screws and Nails

I used only traditional Japanese ties of twisted black hemp to join my first bamboo projects. After hundreds of knots and worn-out hands and shoulders,

If you sink a screw directly into bamboo, the culm cracks and eventually splits. Unlike wood, bamboo doesn't "give." You can use screws successfully by first drilling pilot holes. The trick is to use a drill bit the same diameter as the screw, not smaller as you do with wood. When attaching bamboo to lumber,

drill a pilot hole completely through the two walls and just slightly into the wood. When you drive in the screw, secure it solidly in the lumber. When attaching bamboo to bamboo, rasp the areas where the poles intersect. With the two pieces held in place, drill a pilot hole through the first three walls, then follow with a screw. If the two poles wobble after joining, secure with wire or cordage.

As it enters the hollow internode, a screw point may not easily find its way to the next pilot hole. If it misses, it could create a crack. Prevent this by taking care to drill the pilot holes in a straight line. If your drill has a built-in level, use

it! It also helps to work with two drills—one fitted with the drill bit and the other with a Phillips-head bit—to eliminate the task of changing bits. It's also helpful to keep the poles very stable with your hands or quick-release clamps so they don't shift between the time you drill pilot holes and drive in the screw.

Nails penetrate bamboo easily but are difficult to remove. It's easier and less damaging to remove screws with a reversible power drill than to pull out nails. Nails do, however, work well when attaching bamboo to lumber. For a tight fit, used *ringed nails* (also called *panel nails*). The shafts have raised metal which grabs the fibers of bamboo. Be sure of the nails' placement beforehand; once in, they're difficult to remove. As with screws, you must first drill a pilot hole with a diameter equal to the nail.

Lashing and Binding

For thousands of years, lashing with thin strips of bamboo or twisted plant material has been the most universally embraced technique for joining bamboo. Lashing can be as simple as wrapping and tying two pieces together with cordage, or as complex as securing angled architectural joints that in turn support roof beams and joists. A few simple lashings are all you'll need to get started.

Ringed nails hold bamboo securely.

Top, left to right: square lashing, front crossed lashing, spiral front and back, lashing with spiral, lashing with twisted spiral. Bottom, left to right: traditional Japanese tie, square lashing, basic cross tie, square lashing, traditional Japanese tie.

CORDAGE

Most tying material is made from natural fibers. Coconut fiber, hemp, and sisal binder twine are suitable for outdoor projects. For decorative projects, waxed linen and twisted sea grass work well. Jute is flexible and easy to tie but rots quickly; the rougher strands of hemp and coconut better withstand the stress of sun and rain.

Traditionally, the Japanese use brown hemp twine and black hemp twine, which is made by dyeing the brown twine with carbon black. After several seasons of weathering, the rough fiber twine frays and fades to a silver gray. There is an abundance of alternative lashing materials, including fluorescent masonry cord, bicycle inner tubes, vines, and telephone wire. Creativity rules!

Clockwise, from upper left: dyed black hemp, dyed brown hemp, binder twine, twisted coconut fiber, dyed red hemp, twisted seagrass.

Here are several all-purpose ties, useful for basic attachments. If lashing intrigues you, apply more complex wrapping and knots.

BASIC CROSS TIE

This lashing steadies the two poles by crossing in back and tying in front. Start with 18 inches (45.7 cm) of cordage. Referring to figures 1 through 3, proceed as follows:

1. Fold the cord in half and place diagonally across the intersection, both ends pulled toward the back.

2. In the back, twist the ends tightly together.

3. Bring the two ends back to the front and tie with a square knot.

Figure 1

Figure 2 (back)

Figure 2 (front)

Figure 3

SQUARE LASHING

This is a sturdy lashing, good for structures. Starting with 36 inches (91.4 cm) of cord, refer to figures 1 through 3:

1. Attach one end with a clove hitch. Wrap the cord in a counterclockwise direction, bringing it under, over, under, and over the four poles. Pull tight at each wrap. Complete two or more full rotations.

2. Wrap between the poles, pulling tight and keeping each turn taut.

3. Tie the ends together in a square knot. Cut the ends.

Figure 1

Figure 2

Figure 3

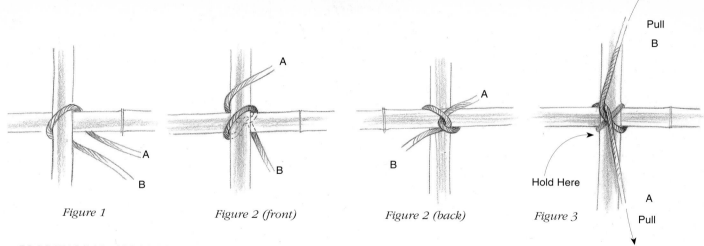

Figure 1

Figure 2 (front)

Figure 2 (back)

Figure 3

Pull
B

Hold Here

A
Pull

TRADITIONAL JAPANESE TIE

This tie is suited for use as a decorative knot, not a strong lashing. It is often tied with two strands of twine (illustrations and directions are for a single strand). Using poles 1 inch (2.5 cm) in diameter, start with a length of 28 inches (70 cm) and refer to figures 1 through 5.

1. Place the twine diagonally across the intersection with both ends toward the back. Adjust the twine so cord A over the pole is 2 inches (5 cm) longer than cord B. Figure 1.

2. In back, cross and tightly twist. Bring the ends forward, the longer one (A) from the upper left corner and the shorter one (B) from the lower right corner. Figure 2, front and back.

3. Cross and tighten at the intersection. Bring cord A up. Keep taut by pinching the twine. Figure 3.

Figure 4

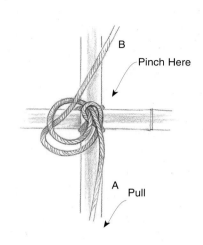

Figure 5

4. Bring A over and under B then up, making a clockwise loop. As it meets the inter section, place A under itself. Figure 4.

5. Twist B under A. Take B clockwise under, around, and through the loop. While pinching with your left hand, use your right hand to pull A to complete and tighten the knot. Figure 5.

6. Clip the ends to about 1 inch (2.5 cm).

USING WIRE

Galvanized steel wire and copper wire work well as lashings if the bamboo is cured and you use flexible wire and snub-nose or needle-nose pliers to achieve a tight twist. If green bamboo is lashed with wire, the joins will loosen when the bamboo dries and shrinks. If the wire isn't flexible enough to wrap tightly around a joint, the bamboo will slip and slide. To get a truly tight joint, rasp the points of intersection, join them with a screw, and stabilize the joint with wire. This works well during project assembly, when adjustments can be made before the final stage of wrapping. If desired, hide screw heads by twisting wire ends into a decorative spiral. Wire can also be lashed and twisted over cordage to stabilize structures.

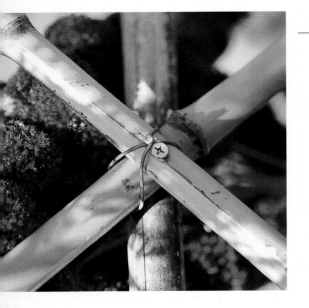

FRONT CROSSED LASHING

1. Start with a piece of wire 18 inches (45.7 cm) long. Bend it

Figure 1

in half and place it behind the horizontal pole. Bring the lower end up and across the other end.

Figure 2 (front)

2. Make a sideways twist and take the ends to the back.

Figure 2 (back)

3. Use pliers to twist the ends together. Cut off the ends.

Figure 3

WIRE CROSS TIE WITH SPIRAL

Follow the directions for the Basic Cross Tie on page 44, and use needle-nose pliers to bend the ends into a spiral. If you're working with thin-gauge wire, twist the ends together then bend them into a spiral.

SPIRAL FRONT AND BACK

1. Start with a 6-inch (15.2 cm) piece of sturdy wire. Use needle-nose pliers to make a 90° bend in the middle. Starting at the bend, work the wire into a spiral.

2. Using a drill bit with a diameter that matches the wire, drill a hole through the poles in the center of the intersection.

3. Insert the straight end of the wire through the holes. In the front, adjust and flatten the spiral to fit the bamboo's curve. In

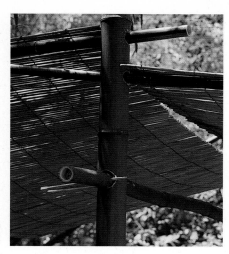

Bamboo joinery techniques were used to make this shade canopy. Design by Harry Abel

back, use needle-nose pliers to pull the straight end tight. Holding the front spiral steady, bend the back end into a spiral.

Joints

Simple joinery is done by inserting the end of a small-diameter bamboo pole into a hole drilled through one or both walls of a pole that's larger in diameter. Start by marking the outline of the smaller piece onto the larger. Select a Forstner or hole-saw bit just slightly smaller than the outline, and drill through the wall. Use a round rasp or Japanese knife to scrape the walls of the holes, customizing

the fit to the incoming pole. You can also drill through the other wall and slide the smaller diameter pole through both walls.

MAKING PEGS

Pegs are useful for stabilizing joints. They're the equivalent of wooden nails and are made from the thick-walled base of a culm. To make pegs, first estimate the peg length required, a length equal to the diameter of the pole plus 1 inch (2.5 cm) on either side. Cut the length from the culm base, split it in half, and use a hammer to knock out the diaphragms. Next, determine the diameter of the peg and make the splints slightly larger. Bracing one end on a work surface, use a single-bevel knife to shave the pulpy inner walls to a tapered end. Be sure to leave the hard outer wall

intact as it is the strongest part of the peg.

To peg a joint, drill a hole through all four walls. Place the peg in the hole so that the outer wall of the peg faces toward one side of the pole, rather than facing up or down. Otherwise the pole may split. Use a hammer to pound in the peg. The fibers of the peg compress between the drilled holes and fit snugly. For even more stabilization, cover the tapered peg with wood glue before driving in. Use a fine-tooth saw to cut off the protruding ends.

FISHMOUTH JOINTS

This term refers to the shape of the carved end of a bamboo pole; it resembles the open mouth of a fish. When placed against the curve of an adjoining pole, it fits closely against it as shown in figure 1 on page 48. To make the fishmouth, estimate how deep the cut-out curve needs to be in order to fit snugly against the adjoining pole. Mark this distance on the opposite side of the pole. With a fine-tooth saw, make short diagonal cuts to form a V, with the point of the V at the mark. To smooth the cuts, rub with a sanding pole made by wrapping a sheet of sandpaper around a pole that's the same diameter as the pole

Fishmouth joint
Figure 1

the fishmouth will join. Leave 6 inches (15.2 cm) unwrapped for a grip, and secure with masking tape. Fit the sanding end up against the cut, and rub the curved edges back and forth until the cut is smooth and fits well with the pole to be joined. As you work on a project, keep several poles wrapped with fine-, medium-, and coarse-grit sandpaper handy.

Bending Bamboo

Sometimes curved culms of bamboo will enhance a project's structure or aesthetics. If you don't have naturally-curved lengths, two bending techniques work well. Both can also be used in reverse to take the curve out of a pole.

FIRE AND WATER

This technique creates gradual, slight curves. The heat of a fire is used to soften the oil and wax in the culm, the fibers are gradually stretched, and water is applied to cool the culm so it holds the bend. First, mark the section of the culm to be curved, and mark the midpoint of the inside curve. Choose a work area with two trees or stationary posts to serve as leverage to make the bend. Build a contained fire nearby and have ready a cloth or large sponge, a large bucket of cold water, and if possible, a friend to help you.

Rotate the area of the curve over the fire until it's hot (but don't scorch it). Quickly position the hot culm between the trees or posts. While your partner stabilizes one end, firmly pull or push your end until you feel a slight give in the culm. Move the warm area back and forth against the post. When you feel the culm has bent to its maximum, stop and hold it firmly in position while your partner drenches the section with cold water. Repeat the process of heating, positioning, bending, and cooling, working alternately to the left and right of the midpoint, until the desired curve is achieved.

KERFS

To create a more extreme curve, you can cut triangular slices called *kerfs* along the inner bend of a culm as shown in figure 2. First, mark a line along the midpoint of the inside curve. Mark and saw out identically sized, equidistant triangular pieces. Using your own strength or working between

Kerfs
Figure 2

fixed posts, bend the culm. If the curve isn't round enough, cut out more kerfs or make the existing ones larger. Maintain the curve by tying or screwing the ends of the pole to other fixed poles.

Flattening

There are two ways to transform bamboo's cylindrical form into a relatively flat surface: one is a Japanese technique and the other, a Latin American technique. Either way, the idea is to break apart some—but not all—of the fibers to allow the culm to open up and spread out.

Japanese Flattening Technique
Figure 3

JAPANESE FLATTENING TECHNIQUE

Split a 2- to 3-inch (5.1 to 7.6 cm) culm in half, and use a hammer to knock out the diaphragms. Lay the culm inner side facing up and clamp it securely to a stable surface. With strong, sure strokes, hammer the chisel straight into the pulp of the diaphragm walls, penetrating and separating the cellulose fibers. Don't fully penetrate to the node, however. Repeat the chisel work at regular intervals along each node as shown in figure 3, with the distance between chisel strokes determined by the desired width of the finished slats. For example, if you want your slats to be 1 inch (2.5 cm) wide, separate your chisel marks by 1 inch or

more. Take care to strike each of the diaphragm walls at slightly altered points across the length of the culm. Otherwise, the fibers may split perfectly along the grain and the culm may break apart. When you finish chiselling, turn the piece over and use the weight of your hands and body to gradually flatten the piece. If necessary, make additional chisel strokes to get a flatter surface. Lay pieces of lumber across it, and clamp it down overnight. Screw or nail in place.

ESTERILLA

Derived from the Latin American Spanish word meaning mat or matting, this technique flattens the culm by creating an accordion-like series of splits.

Bamboo esterilla is used in Central America as matting, wall paneling, and table veneers. Use the mid-section of the culm, where the diameter is more or less constant.

Begin by splitting a pole in half, and use a hammer to knock out the diaphragms. From one side, measure 3/16 inch (4.8 mm); use a knife and mallet or the tapping technique to embed the blade. Twist the blade, moving it down to create a split about three-fourths of the way down. Flip to the other end, measure 3/8 inch (9.5 mm), and split along the grain the same distance as your first split. Flip again, measure 3/8 inch (9.5 mm) from the previous split, and split. Continue back and forth until you reach the other side.

Gently spread open the split pole as shown in figure 1 on page 50.

SPLITS

OPENED

Esterilla Flattening Technique
Figure 3

With a knife that has a long blade, cut off the diaphragms of the inner wall. Run the piece through a planer, shaving off the rough inner pith to achieve a flat, uniform surface. The esterilla is now ready for its intended purpose. Refer to the Side Table with Bamboo Esterilla on page 74 for one example.

Finishing and Preserving

Left untreated, bamboo weathers quickly when exposed to the elements. Finishing its outer surface preserves the structure while giving it richness and sheen.

SWEATING

The process of heating fresh or cured bamboo over a fire or flame is called *sweating*. Heat causes the oil and wax in the culm to rise to the surface and bead up like sweat. By removing excess secretions and rubbing with a cloth, you can give

the outer layer a hard surface and beautiful sheen.

First, wash and scrub off any grime from the lengths you'll sweat, because dirt gums up the released oil and is difficult to remove. Make a fire in a safe, outdoor enclosure such as a barbecue grill or a galvanized metal tub with holes in its bottom and sides, set on bricks. The fire should burn hot and fast, so have a good supply of kindling and fuel nearby. If an outdoor fire isn't an option or you want to sweat just a few poles, you can use a stove or handheld propane torch. For large projects, a roofing torch that runs off a five-gallon (19 L) tank of propane works well. Scraps of bamboo make good fuel, but be careful to use only pieces that are completely split open. Lengths of bamboo with closed internodes can explode in fires and send fragments flying. One Malayan tale claims that such a loud and sudden

noise is what gave the plant its name: "bam-BOO!"

Hold a portion of the pole 6 to 12 inches (15.2 to 30.5 cm) above the fire, moving it in sweeping circular motions.

When you see wax beading up on the surface, let it build up, then remove the culm from the fire and use a soft cloth to quickly wipe off the oil and buff the surface. Return an overlapping section of the culm to the fire and repeat the rotating, wiping, and buffing process. The green culm will change to a shiny tan. Unless you want to create dark areas, be careful not to hold one area of the culm over the flame too long, or the oil will burn and scorch the surface.

POLISHING

A quick way to make a dull culm shine or to remove the chalky-white deposit on the surface of some bamboos is to use fine-grade steel wool. Wet the bamboo or dip the steel wool in water, then rub the pad up and down. Work in sections until the surface achieves a luster.

SCRAPING AND DYEING

If you want a refined, fresh-looking surface, you can scrape away the outer layer of green or cured bamboo to expose the delicate longitudinal strands of fiber. Hold the top of the culm at an angle with its base resting on a wooden surface. Hold the knife blade on an angle and move it up and down in a repetitive movement as the delicate shavings fall. Patience and strong arm muscles can give your bamboo a beautiful surface. For color, use wood stains, wood dyes, or fabric dyes formulated for cellulose fibers. Apply one or more coats.

SEALERS

For bamboo that will remain indoors, sealers such as polyurethane will preserve and finish your work. Outdoors however, sun, heat, and moisture will cause the sealer to peel away after a year or so of exposure. The best way to keep outdoor bamboo structures looking their best is annual maintenance.

STAINS

Exposed to the elements, the surface layer of bamboo eventually breaks down and becomes pitted. Mold and mildew may accumulate and lead to decay. On the good side, the rough surface will accept finishes which otherwise won't adhere to bamboo's waxy silicone surface.

Stains give cured bamboo a new lease on life. Wash cured and weathered bamboo to remove dirt and mildew, and let dry. Use a soft cloth to rub the stain onto the culm, starting at the base and moving up. The stain adheres to the surface, rather than penetrating it, and a second coat may be necessary to achieve the full depth of color.

Gel stains are non-drip and work well on scraped bamboo. "Golden oak" tones give a natural look.

ANNUAL MAINTENANCE

It's a good practice to make an annual inspection of outdoor bamboo structures. Fall is a natural time to do this, after vegetation has died down but the weather is still mild. Remove any vines, vegetation, and debris lodged around or between the poles. Wash the structure by thoroughly wetting it with a garden hose, then spraying with a nonsudsing cleanser. Give the cleanser time to act, then loosen the grime with a scrub brush, paying special attention to joints and intersections. Rinse thoroughly with fresh water and let dry.

Chapter 6

Bamboo in the Garden

Bamboo bridge by Kat Semrau. Photo by Rita Randolph

JUST ABOUT ANY GARDEN CAN BENEFIT FROM STRUCTURES MADE OF HARVESTED BAMBOO. UPRIGHT POLES WITH THEIR TOP ENDS CUT JUST ABOVE THE NODE MAKE FRIENDLY RESTING POSTS FOR THE BIRDS WHO CONTROL THE INSECT POPULATION. BAMBOO TRELLISES GIVE GRACEFUL SUPPORT TO FLOWERING AND FRUITING VINES.

IF YOU GROW YOUR own bamboo, the garden is a good place to use your harvested culms. Keep assorted lengths stored near your garden; they'll come in handy for all sorts of garden projects and puttering.

Stabilizing Poles

You'll find several techniques handy for stabilizing freestanding bamboo structures. To set poles in the ground, cut the bottom end on the diagonal. Use a hammer to pound a piece of rebar through the base 18 to 24 inches (45.7 to 61 cm) into the pole. Knock out the nodes and ream out as much membrane as possible. This will give the dirt a place to go when the pole is hammered into the ground. Use a rubber mallet to drive in the pole; if you use a hammer with a steel head, place a wood scrap over the top of the pole to protect it.

You can also use rebar fitted with bamboo sleeves. This technique keeps the poles off the ground, lengthens the life of the structure, and allows you to remove it for storage or repair. Use a hammer to pound rebar 3 to 5 feet (0.9 to 1.5 m) through the pole, knocking out the diaphragms. This creates a hollow tube or sleeve. Drive a 3- to 5-foot (0.9 to 1.5 m) length of rebar 18 to 24 inches (45.7 to 61 cm) into the ground or deep enough to assure stability. Then slide the hollowed bamboo pole over the rebar. To decrease rot, place bricks on either side of the rebar to keep the bamboo off the ground.

Clockwise from upper left:
Bamboo trellis with copper tubing.
DESIGN AND PHOTO BY PETER GALLAGHER

This bamboo trellis supports growing vines.

A length of bamboo is combined with straw to create a birdhouse.

Bamboo provides support for bushes and shrubs.

Bamboo adds a calming presence to a water garden.

Lightning Trellis

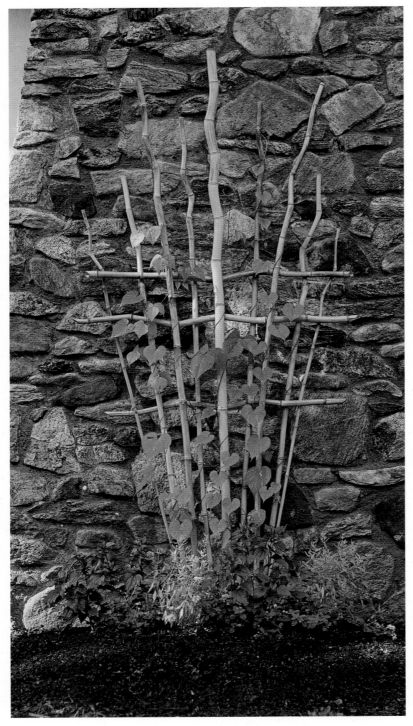

THANKS TO THE ZIGZAG SHAPE OF GOLDEN BAMBOO (*PHYLLOSTACHYS AUREA*), THE HORIZONTAL POLES OF THIS DELIGHTFUL TRELLIS REST IN THE CROOKS OF THE VERTICAL POLES. THE OVER-AND-UNDER PATTERN GIVES DIMENSION AND ADDITIONAL VISUAL INTEREST. ARRANGE AND ADJUST THE FULL-LENGTH POLES ON A FLAT SURFACE TO GET THE BEST FIT BEFORE YOU CUT.

Instructions

1 Refer to tools and materials listed on page 56. Select one of the nine vertical poles to serve as the center pole. It should be strong and have an appealing zigzag shape. Use the hammer and rebar to knock out the lower 3 feet of diaphragms.

2 On a large, flat surface, lay out the vertical poles in a fan shape with the zigzag ends up. Start by placing the center pole in the middle and work outward, bringing the poles close together at the base, and fanning them out to 60 inches at the widest part. Position and adjust the poles to get a good fit.

3 One by one, lay the vertical poles over and under the horizontals where poles and zigzags fit together. Work to create a surface with dimension. See figures 1 and 2.

4 When you're satisfied with your layout, use the saw to cut the poles to their finished lengths.

5 Where the poles intersect, drill pilot holes and attach with the drywall screws on one side, then turn over the structure and screw from that side.

6 At the installation site, mark the center point, and hammer in the rebar 18 to 24 inches deep. Place a brick on each side of the rebar.

Figure 1

Figure 2

Materials

5 1¼-inch drywall screws

9 1⅝-inch drywall screws

9 2¼-inch galvanized decking screws

Rebar, 4 feet long, to serve as center pole support

2 bricks

Tools and Supplies

Fine-tooth saw

Power drill and assorted drill bits

Sanding sponge

Wire (optional)

Nails (optional)

Cutting List

Description	Qty.	Material	Dimensions
Verticals	9	Golden bamboo poles, zigzag-shape	1" to 1¾" x 6' to 8'6"
Horizontals	3	Golden bamboo poles, zigzag shape	1¼" x 38", 54", and 62" respectively

Metric Equivalents

1"	2.5 cm	1¾"	4.4 cm	38"	96.5 cm	3'	0.9 m		
1¼"	3.2 cm	2¼"	5.7 cm	54"	137.2 cm	4'	1.2 m		
1⅝"	4.1 cm	18"	45.7 cm	60"	1.5 m	6'	1.8 m		
		24"	61 cm	62"	57.5 cm	8'6"	2.5 m		

7 Hold the trellis over the rebar and slide the central pole over the rebar, resting the base of the poles on the bricks. If more stability is required, use the wire and nails to attach the trellis to the wall.

Water Dipper

with assistance from Dong Lingen and Reed Hamilton

BAMBOO WATER DIPPERS ARE PLACED AT THE ENTRANCE OF MANY JAPANESE TEMPLES AS AN INVITATION FOR VISITORS TO TAKE PART IN RITUAL CLEANSING. THE DESIGN OF THIS DIPPER IS APPEALINGLY SIMPLE AND FUNCTIONAL AND CAN BE CRAFTED FROM SMALLER-DIAMETER BAMBOO AS WELL. TO GIVE IT A LONGER LIFE, REST THE DIPPER UPSIDE DOWN TO THOROUGHLY DRAIN THE WATER WHEN IT'S NOT IN USE.

Materials

1 bamboo splint, 21 inches long, 1¼ inches wide, ¼-inch thick, to serve as the handle

1 piece of bamboo, 4 to 5 inches long, 3½ inches in diameter, to serve as the bowl*

*This piece should have 1 node and ⅜-inch thick walls

Metric Equivalents

¹³/₆₄"	5.2 mm	1⅛"	2.8 cm
¼"	6 mm	1¼"	3.2 cm
½"	1.3 cm	2"	5.1 cm
⅜"	9.5 mm	3½"	8.9 cm
⅝"	1.6 cm	4"	10.2 cm
¾"	1.9 cm	5"	12.7 cm
¹³/₁₆"	5.2 mm	7"	17.8 cm
1"	2.5 cm	21"	53.3 cm

Tools and Supplies

Quick-release clamps

Power drill (or drill press) with 1⅛-inch Forstner drill bit and ¹³/₆₄-inch drill bit

Fine-tip permanent marker

Ruler or straightedge

Fine-tooth saw

Japanese cutting knife

Shaver

Sanding sponge

Sandpaper, medium grit

Pure beeswax candle

Matches or lighter

Piece of flexible, waxed cardboard, 1 x 2 inches

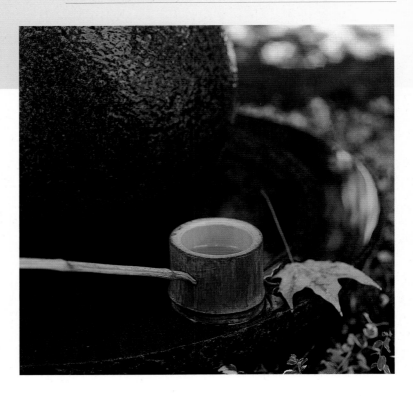

Instructions

1 To make the handle, mount the Forstner bit in the drill or drill press. Hold the bit to one end of the splint and create a curve on the end. This is most easily accomplished with a drill press, but if you use a handheld drill, use a vise or clamp the piece securely to your work surface, and hold the drill steady as you drive it through the splint.

2 Starting at the curved end of the splint, mark a paddle-shape outline that's ¾ inch wide at one end and narrows to a ½-inch neck. Along each side of the handle, use the ruler to make straight lines that start out ½ inch apart at the neck, then very gradually widen so that at a point 7 inches up from the neck, they are ⅝ inch apart. Maintain the ⅝-inch width to the end of the handle.

3 Referring to figure 1, use the knife to carve out the base, neck, and sides of the handle along the lines. Use the knife or shaver to remove splintery edges,

and smooth with the sanding sponge.

4 Now you'll make the bowl. From the length of bamboo 3-½ inches in diameter, use the fine-tooth saw to cut just below the node. Make sure you cut evenly across so the bowl sits level. If you have doubts about making an even cut, mark a straight line around the perimeter to guide you as you cut.

5 Use the knife to round off the inner and outer rim. Follow with the sanding sponge, smoothing the rim of the base.

6 As shown in figure 2, decide the side of the bowl to which you wish to attach the handle. Trace the curved outline of the handle base onto this area, 1 inch below the rim.

7 Use a vise or clamp the bowl so it sits open end up. Holding the ¹³⁄₁₆-inch drill bit at the angle desired for the handle, drill in between the outline, the bit pointed down and in. Drill a series of holes overlapping each other along the curve of the outline until the opening has been made.

8 Use the knife to cut out the pieces of bamboo that remain between the holes. Then fold up a piece of the sandpaper and run it back and forth through the opening until the inside edges are even.

9 Insert the base of the handle in the hole. The joint should be snug with the handle fitting securely in the bowl. If the handle won't quite go in, mark and rework until the fit is satisfactory. You can also trim the outer wall of the handle to decrease its width.

10 Now you'll attach the handle to the bowl. From the inside of the bowl, hold the waxed cardboard securely over the opening. Now light the beeswax candle. Let the hot wax drip into the curved opening until it's half full. Blow out the candle and quickly insert the handle into the bowl. Hold the cardboard tight so the wax doesn't ooze into the bowl and hold the handle and bowl steady for several minutes until the wax cools. Place in a cool place until the wax thoroughly hardens.

11 Use the knife to remove excess wax from the inside or the outside of the bowl.

Figure 1

Figure 2

Low Garden Trellis

Designer, Anita Matos

THIS SEE-THROUGH, DIAMOND-PATTERN TRELLIS IS A NATURAL FOR CUCUMBERS, GOURDS, AND OTHER CLIMBING VINES. AS SHOWN HERE, IT CAN UNIFY A GARDEN GONE GLORIOUSLY WILD WITH CULTIVATED AND "VOLUNTEER" FLOWERS AND VEGETABLES. A THIRD POST IN THE TRELLIS PROVIDES EXTRA SUPPORT FOR WEIGHTY PRODUCE.

Materials

2 lengths of rebar, each 4 feet long

60 drywall screws in lengths of 1, 1¼, 1⅝, and 2 inches

Galvanized wire, 16-gauge

Copper wire

Binder twine

Tools and Supplies

Hammer

Measuring tape

Power drill and assorted drill bits

Pliers

Wire cutters

Cutting List

Description	Qty.	Material	Dimensions
Horizontals	3	Bamboo poles	1½" x 7'
Posts	2	Bamboo poles	1½" x 4'
Long diagonals	11	Bamboo poles	½" to 1" x 6'
Short diagonals	3	Bamboo poles	½" to 1" x 4'

Instructions

1 Working with the two 4-foot poles that will serve as posts, use the hammer and a piece of the rebar to knock out the lower 3 feet of diaphragms.

2 On a large, flat surface, lay out the posts 6 feet apart. Place the horizontal stringers on top, one 3 inches below the top of the post, one 3 inches above the base, and one in the middle. Drill pilot holes at the intersections of posts and stringers, and drive in the 2-inch screws.

3 Mark the center of each horizontal. You'll use the marks as reference points when placing the diagonals to form the diamond pattern.

4 Place five of the long diagonals and one of the short diagonals on the frame, all leaning in the same direction. Center the diagonals, drill pilot holes, and use the appropriate screws to secure the ends to the frame.

5 Turn over the trellis assembly. Lay out the remaining diagonal poles in the opposite direction. Voilá, diamond patterns! Adjust, drill pilot holes, and screw at intersections.

6 To install the trellis, mark the placement of the two end posts. Use the hammer to pound the rebar into the ground at those points. Lift up the trellis and slide the posts over the rebar.

7 Use the wire or cordage to reinforce intersections that need extra support.

Metric Equivalents

½"	1.3 cm
1"	2.5 cm
1¼"	3.2 cm
1½"	3.8 cm
1⅝"	4.1 cm
2"	5.1 cm
3"	7.6 cm
3'	0.9 m
4'	1.2 m
6'	1.8 m
7'	2.1 m

Plant Stakes and Markers

RATHER THAN HUNTING FOR FALLEN BRANCHES TO MARK SEEDLINGS OR TIE UP LEANING PLANTS, YOU CAN EASILY SPLIT BAMBOO INTO HANDY STAKES AND MARKERS. THEY'RE AN EXCELLENT USE FOR THOSE IRREGULAR SCRAPS AND ARE LONG-LASTING. IN THE FALL, PULL THEM OUT OF THE GROUND, SHAKE OFF THE DIRT, AND STORE FOR USE IN THE SPRING. BUNDLES OF SMOOTH STAKES TIED WITH TWINE MAKE GREAT GIFTS FOR YOUR GARDENING FRIENDS!

Materials, Tools, and Supplies

5 or more bamboo poles, 12 to 60 inches long, 1¼ inches or greater in diameter

Splitting knife and mallet, or a four-way splitter

Hammer

Knife or shaver

Sanding sponge

Fine-tip permanent markers (optional)

Metric Equivalents

1¼"	3.2 cm
12"	30.5 cm
60"	1.5 m

Instructions

1 Use the splitting knife and mallet to split the lengths of bamboo in half, then hammer out the diaphrams. Split each length in half again, forming a total of four splits per pole. If you're using a four-way splitter, split the lengths, then remove the quartered node membranes with a knife.

2 Use the shaver or knife to smooth the edges and eliminate splinters.

3 Rub the sanding sponge up and down the splint several times to smooth the edges and the inner wall.

4 If desired, use the fine-tip permanent marker to write plant identifications or other information directly on the bamboo. Color them, too, if you're feeling festive!

Dripper and Deer Scare

Designer, Harry Abel

THIS PAIR OF WATER GARDEN ACCESSORIES WORKS TOGETHER TO CREATE A SIMPLE AND ENCHANTING DISPLAY OF MOVEMENT AND SOUND. A SLIGHT STREAM OF WATER FLOWS FROM THE *SOZU*, OR DRIPPER, INTO THE WAITING END OF THE *SHISHI ODOSHI*, THE DEER SCARE. THE SCARE'S HOLLOW MOUTH SLOWLY FILLS. AFTER THE WEIGHT OF THE WATER CAUSES THE SCARE TO TIP AND EMPTY, THE BASE END DROPS QUICKLY TO STRIKE A ROCK, PRODUCING A HOLLOW CLACKING SOUND. IN THE JAPANESE CULTURE, THE SLOW MOVEMENT OF WATER, THE ACTIONS OF FILLING AND EMPTYING, AND THE PUNCTUATION OF SUDDEN SOUND ARE COUNTERPOINTS TO THE STILLNESS OF THE GARDEN AND MARKERS OF THE PASSAGE OF TIME.

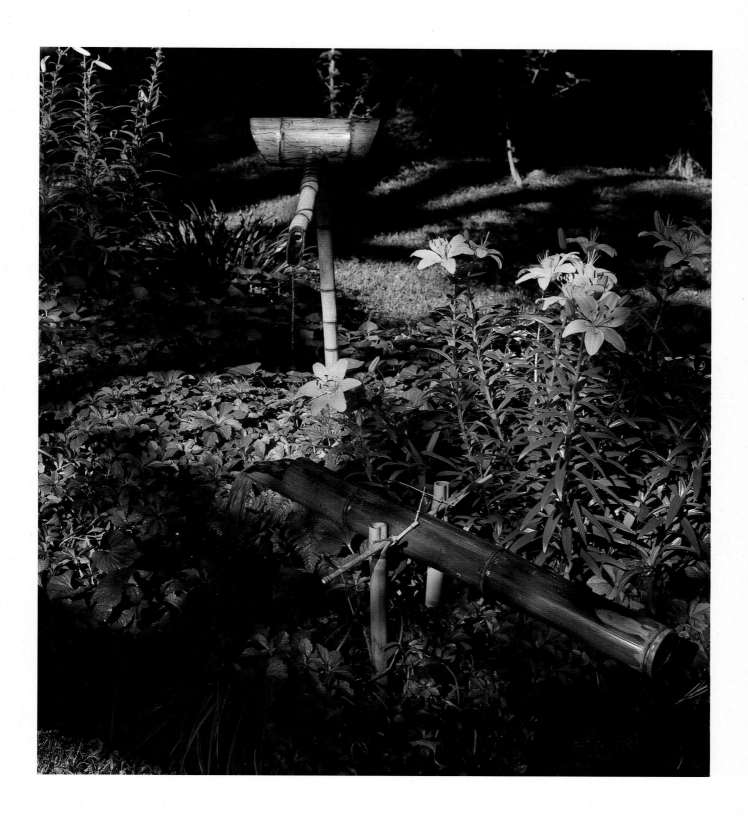

Materials

1 36-inch piece of ¼-inch rebar, to serve as support for dripper uptake pole

2 24-inch pieces of ½-inch rebar, to serve as supports for scare posts

Polyethylene tubing, 72 inches long, ½-inch outside diameter

1 smooth, curved, dark rock to serve as the sounding rock

Water source such as a pond or basin

Recirculating water pump

Tools and Supplies

Measuring tape

Fine-tooth saw

Vise or clamps

Power drill with 1½-inch Forstner bit and ½-inch brad-point bit

Hammer

Sanding sponge

Garden hose or spigot

Cutting List

Description	Qty.	Material	Dimensions
Dripper uptake pole	1	Bamboo pole	1½" x 36"
Dripper head	1	Bamboo pole with nodes at each end*	5" to 6" X 8½"
Dripper spout	1	Bamboo pole	1½" x 12"
Scare water collector	1	Bamboo pole with 4 or 5 nodes	3" x 36"
Scare pivot bar	1	Bamboo pole	½" x 6"
Pivot bar posts	2	Bamboo lengths with branches**	1½" x 24"

*This piece can be taken from the base of a culm at a point where the nodes are close and the culm diameter is widest.

**Cut the top of these pieces just above the node, and trim the two branches that extend from the node to 6 inches.

Instructions

Making the Dripper

1 Use the rebar and mallet to knock out all the diaphragms of the 36-inch uptake pole. Ream as close to the walls as possible to allow room for both the plastic tube and the 24-inch length of rebar. Determine which end of the uptake pole is closest to 1½ inches in diameter, and use the saw to cut that end at a 45° angle.

2 Secure the head of the dripper in a vise or clamp it to a worktable. Use the 1½-inch Forstner bit to drill a hole through one wall of the dripper midway between the two nodes. Turn the head 90°, and drill another hole midway between the nodes. The holes will receive the uptake pole and the spout.

3 Now you'll make the spout. Take the 12-inch length of bamboo and determine which end fits most tightly into the head.

Use the saw to cut this end at a 45° angle with the opening facing down. Cut the mouth end at a 45° angle with the opening facing up.

4 Determine the placement of the dripper in relation to the deer scare. At the site intended for the dripper uptake pole, hammer the ¼-inch rebar 18 inches into the ground.

5 Refer to figure 1. Snake the plastic tubing through the base of the uptake pole until it

Spout

Uptake
pole

Tube,
attached to
pump

Rebar

Figure 1

emerges from the angle-cut top. Maneuver the tube through the holes in the head, then fix the head on the uptake pole. Slide the end of the tubing through the spout, stopping before it is visible. Secure the angled end of the spout into the head, making sure the angled cuts fit together and the tube is not crimped.

6 Insert the other end of the plastic tube into the water pump.

Making the Deer Scare

1 First, you'll make the water container. Select the base end of the 36-inch-long, 3-inch-wide pole. This will strike the sounding rock, and the opposing end will collect water. Just below the node, use the saw to cut the base straight across, and cut the mouth end at a 45° angle. Smooth the edges with the sanding sponge.

2 Use your thumb and forefinger to lightly grasp the pole at the center of the middle node. Drip water from the garden hose into the mouth of the scare. When the mouth is full, the tube should tip in your fingers, spilling the water. If it doesn't tip, test for the pivot point by sliding your fingers up and down the length. Mark this pivot point on both sides of the pole.

3 Use the brad-point bit to drill through the mark on one wall into the mark on the other wall. Just as the point emerges, stop. Finish drilling the hole from the outside wall to prevent splitting and feathering. Sand the edges. Insert the wide pivot bar through the two holes and center it.

Installing the Deer Scare and Dripper

1 Use the hammer and rebar to knock out the diaphragms from the base up to, but not including, the top node of the two 24-inch post lengths.

2 Referring to figure 2, determine the placement of the scare in relation to the dripper. Place the two 24-inch pieces of rebar 5 inches apart, and hammer each one 12 inches into the ground.

3 Slip the posts over the rebar, aligning the base of the branches so they're equal in height.

4 Place the scare and pivot bar between the Y of the branches so the pole rocks easily back and forth. Trim the branches to a pleasing form.

5 Place the sounding rock so that when the water empties and the pole swings to rebalance, the returning end strikes the rock squarely. If necessary, adjust the water pressure and the placement of the sounding rock.

Sounding
rock

Figure 2

Metric Equivalents

¼"	6 mm	8½"	21.6 cm
½"	1.3 cm	12"	30.5 cm
1½"	3.8 cm	18"	45.7 cm
3"	7.6 cm	24"	61 cm
5"	12.7 cm	36"	91.4 cm
6"	15.2 cm	72"	1.8 m

Chapter 7

Furnishings and Accessories

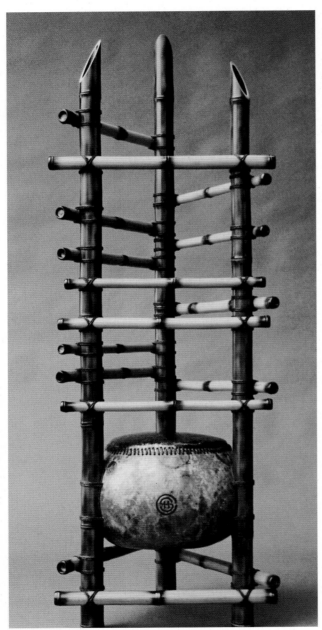

Cal Hashimoto, *Spirit Vessel* - Series V, 1997, 44 x 15 x 15 in. (1.1 m x 38 x 38 cm); bamboo, gourd, brass, twine.
PHOTO BY CAL HASHIMOTO

Right: Cal Hashimoto, *Call to Center* - Series II, 1998, 90 x 26 x 5 in. (2.3 m x 66 x 12.7 cm); bamboo, brass gong, twine.
PHOTO BY CAL HASHIMOTO

Bamboo is as functional and pleasing inside the house as it is outside, and handcrafted bamboo objects lend a natural, outdoor feeling. Its strength and smooth, cylindrical surface makes it a versatile material with which to craft tables, screens, benches, and other furnishings. If you combine bamboo with milled lumber or other materials, the textural contrast adds richness and dimension to the final product.

In addition to furniture, it's easy to make simple bamboo objects to add a serene feeling to any room. Bamboo vases hung on walls or grouped on a tabletop make charming displays for hand-picked flowers. If you're entertaining guests, you can offer food in sushi trays and drinks in bamboo cups. Best of all, a few hand tools and bamboo poles are all you need to make these lovely accessories.

Clockwise from top: Bamboo screens are framed with Western red cedar.
Design and photo by Gordon Powell

Bamboo torchiére lamp designed by John Scheer.

A sunroom at the Bamboo Farm in Savannah, Georgia is faced with split bamboo.

Barbara Schindler, *Harmony Hanger,* 2000, 24 x 24 x 2 in. (61 x 61 x 5 cm); bamboo, hemp twine, beads, amber and turquoise stones.

69

Cedar and Bamboo Inlay Table

with the assistance of Randall Ray

C RAFTSPEOPLE KNOW THAT ELEGANCE IS OFTEN ABOUT RESTRAINT, AND CHOOSE TO HIGHLIGHT ONLY ONE OR TWO DECORATIVE FEATURES IN THEIR WORK. THIS TABLE IS CONSTRUCTED OF BEAUTIFULLY FINISHED CEDAR AND MITERED INLAY MADE FROM BLACK BAMBOO (*PHYLLOSTACYS NIGRA*).

Tools and Supplies

Tape measure

Miter saw

Power drill with ⅜-inch and 1/16-inch bits, and ⅜-inch router bit

30 galvanized decking screws, 2 inches long

Paste construction adhesive

Circular saw or table saw

Wood glue

Fine sandpaper

28 3-inch wood screws

30 2-inch wood screws

12 tapered wood plugs

2 paintbrushes

Clear wood sealer

Wood stain, brownish black color

Fine-tooth saw

Splitting knife

Hammer

56 1⅝-inch ringed nails

Cutting List

Description	Qty.	Material	Dimensions
Frame	1	Western red cedar 4 x 4	8'
Frame and legs	1	Western red cedar 4 x 4	10'
Cleats	-	Pressure-treated 1 x 2 lumber	13' total
Tabletop support	1	Pressure-treated plywood	¾" x 18" x 24"
Inlay	10	Black bamboo poles	1½" x 4' to 5'

Metric Equivalents

1/16"	1.6 mm	16"	40.6 cm
3/16"	4.8 mm	17"	43.2 cm
¾"	1.9 cm	18"	45.7 cm
⅜"	9.5 mm	24"	61 cm
1½"	3.8 cm	37⅝"	95.6 cm
1¾"	4.4 cm	41¼"	104.7 cm
1⅝"	4.1 cm	48"	121.9 cm
2"	5.1 cm	4'	1.2 m
3"	7.6 cm	5'	1.5 m
8"	20.3 cm	8'	2.4 m
13¾"	34.9 cm	10'	3 m
14"	35.6 cm	13'	3.9 m
15¾"	40 cm		

Figure 1

Instructions

Making the Table Frame

1 Refer to figure 1. Cut the 8-foot cedar 4 x 4 in half at a 45° mitered angle, each with a 48-inch long point. From the 10-foot cedar 4 x 4, cut two lengths at a 45° angle, each with a 24-inch long point.

2 Assemble the cedar pieces on the work surface, and use the ⅜-inch drill bit to predrill three ¾-inch-deep pilot holes in each corner (see fig. 2). Use the ¹⁄₁₆-inch bit to drill 2 inches into the center of each hole.

3 Dust off the shavings. Apply the adhesive to the ends of the cut pieces, assemble, and screw together with the 2-inch screws, countersinking ⅜ inch. Allow the glue to dry.

4 From the remaining 4 x 4 pieces, square-cut four legs, each 15¾ inches long. Rabbet the top end of each leg, as shown in figure 3. Cut and remove the three sections, leaving a tenon that fits into the corners of the frame from step 3.

5 Place the frame face down as shown in figure 4. Apply adhesive to the inside of the tenon. Set the tenons of the legs into the inside corners of the frame with the other portions of the leg resting on top of the frame. Drill a ¹⁄₁₆-inch pilot hole 2 inches deep into the outside of each tenon. Drive in the 3-inch screws. Let dry. Sand or chisel off the excess glue.

6 Lay the table on its side, and pencil a line inside the frame 1½ inches from the top edge of the frame. From the 1 x 2 boards, cut two pieces 37½ inches long and two pieces 13½ inches long.

7 Predrill ³⁄₁₆-inch pilot holes 2 inches deep every 8 inches along the line you've drawn.

Figure 2

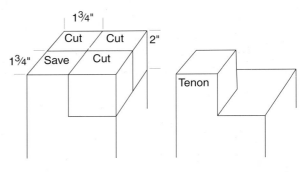

Figure 3

Referring to figure 4, position the cleats you cut in step 6 between the legs, along the 1½-inch depth line. Screw in place with the 3-inch wood screws.

8 Place the table upside down. Cut three 2 x 2 boards to a 15½-inch length. Referring to figure 5, set these stretchers across the width of the table, between the cleats. Drill two ⅟₁₆-inch pilot holes on an angle, then use the 2-inch screws to toenail the ends of the stretchers onto the cleats.

9 Place the table upright. Measure and cut the ¾-inch plywood to 17 x 41 inches. After applying the wood glue to the top edge of the cleats, center the plywood and drop it into the table. Secure the plywood to the cleats with the 2-inch wood screws, three along the long sides, and three along the short sides.

10 Use the ⅜-inch router bit in the drill to rout out the outside top edge of the frame and the base of the legs, creating a pleasing curve. Fill the screw holes with the wood plugs.

11 Clean off any dust or shavings, and use the paintbrush to apply two coats of clear wood sealer to the cedar. Coat the plywood with the stain. Let dry.

Making the Tabletop Inlay

1 Brush the bamboo with the stain and let dry.

2 Use the tape measure and chalk to divide the plywood seat into four equal sections, as shown in figure 6.

3 Use the fine-tooth saw to cut the bamboo into 10 16-inch lengths. Split the bamboo in half lengthwise, creating 20 pieces. Arrange the halves on the seat in a herringbone pattern, as shown in figure 7, adjusting them to get a close fit and an even surface. Cut and split the remaining bamboo, and use it to complete the surface pattern.

4 Use the handsaw and miter box to cut the bamboo at a 45° angle. Work carefully and precisely to achieve a good fit.

5 Lightly sand the ends of the bamboo, and brush the surface and ends with the stain. Allow to dry, then lay the pieces back into the table. Drill a pilot hole through the end of each bamboo piece and into the plywood base. Hammer a ringed nail through the holes. Repeat with each piece until the inlay is complete.

Figure 4

Figure 5

Figure 6

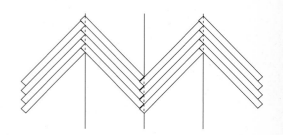

Figure 7

Side Table with Bamboo Esterilla

Designer, Francisco Plaza

ESTERILLA, A LATIN-AMERICAN SPANISH IDIOM FOR MAT OR MATTING, GIVES TEXTURE AND GRAPHIC INTEREST TO THE TOP OF THIS TABLE, A CIRCLE WITHIN A TRIANGLE. PEGS AND TRADITIONAL JOINERY TECHNIQUES ADD TO THE FINE CRAFTSMANSHIP OF THE TABLE.

Materials

16 drywall screws, 1¼ inches long	Fine-tooth saw
Furniture tacks, ½ inch long	Belt sander with sandpaper, #40 grit
Wood glue	3 bungee cords
Brad nails, ½ inch long	Miter box
Wood stain, brownish black color	Round wood rasp

Tools and Supplies

Source of flame	Japanese knife
Straightedge ruler, 36 inches long	Hammer
Pencil	Piece of string, 14 inches long
Jigsaw	Splitting knife with long blade
Circular saw	Mallet with plastic or wooden head
Power drill with 1⅛-inch hole-saw bit, ¼-inch-diameter brad-point bit 10 inches long, and bit with a diameter that matches the furniture tacks	Wood plane
	Clean cloth
	Paintbrush
	Sanding poles (see page 46) with medium and coarse sandpaper

Cutting List

Description	Qty.	Material	Dimensions
Legs	3	Bamboo lengths	3½" x 20"
Rails	6	Bamboo lengths	1¼" x 21"
Tension braces	6	Bamboo lengths	1" to 1¼" x 16"
Esterilla	3	Bamboo lengths*	3½" x 20"
Splint for tabletop trim	1	Bamboo length**	2" x 6'
Tabletop and leg attachments	6	Bamboo pegs	Diameter tapered from ¼" to ⅜", 6½" long
Rail and brace attachments	12	Bamboo pegs	Diameter tapered from ¼" to ⅜", 4" long
Jig	2	Scrap plywood	¾" x 24" x 24"
Tabletop base	1	Plywood	1" x 22" x 22"
Jig supports	6	2 x 2 lumber	12" long
Flattening strips	5	Scrap wood	¼" x 1" x 22"

*with a consistent wall thickness, taken from the midsection of the culm

**taken from upper midsection of a culm

Instructions

Preparing the Bamboo

1 Sweat and scorch the bamboo to produce a rich, brown sheen (see page 50).

Making the Jig

1 Referring to figure 1, use the ruler and pencil to draw an equilateral triangle with 16-inch sides onto one piece of scrap plywood.

2 Place the 20-inch legs top end down at the points of the plywood triangle, and trace around the legs. Use the ruler to draw lines connecting the outsides of the circles.

3 Use the circular saw to cut the straight edges of the plywood, and the jigsaw to cut out the space for the legs. Transfer this shape onto the other piece of scrap plywood, and cut it out.

4 Arrange the pieces of 2 x 2 lumber between the plywood forms as shown in figure 2, placing them toward the center so they won't get in the way of the legs.

5 Use the power drill and screws to attach the supports to the top and bottom plywood forms.

Making the Legs and Table

1 Use the fine-tooth saw to cut each 20-inch leg length just above the top nodes, straight across, being careful not to feather the outer wall. From the top, measure down 18 inches and cut straight across so the legs will sit flat on the floor. Use the belt sander to smooth the ends.

2 Place the legs in the jig, and wrap the bungee cords around them to keep them in place. Place the tabletop onto the plywood surface, setting it between the tops of the legs. When centered, there should be a $\frac{5}{8}$-inch space between the legs and the edges of the tabletop to account for the trim that you'll add later.

3 Refer to figure 3. With the table legs still held in place by the bungee cords, measure and mark 1 inch down from the top of each leg. This indicates the top of the rails that will hold the tabletop. Then starting from the base of the legs, measure up 4 inches to mark the bottom of the lower rails.

4 Use the miter box and fine-tooth saw to cut a 60° angle at each end of the 21-inch rails.

5 Place the rails between the legs and check for fit. Use the

Cut out

Tabletop

1½" Plywood jig

2 x 2 supports

3½"

16"

24"

Plywood jig

Figure 1

Figure 2

marker to outline the circumference of the rails onto the legs. At the top of the legs, outline each rail just below the mark. At the botttom of the legs, outline each support just above the mark.

6 Remove the bungee cords and the legs from the jig. Clamp each leg to the worktable, and use the drill with the 1⅛-inch hole-saw bit to make the holes in the legs. Take care to drill at the proper angle with the top and bottom holes in alignment.

7 Now you'll prepare the legs to be joined to the rails. Use the Japanese knife and round rasp to scrape the sides of each hole, customizing its shape to accommodate the end of the incoming rail and achieve a good fit. Reassemble the legs around the jig, insert the rails, and secure with the bungee cords.

8 Peg together the legs and rails, as shown in figure 4. Use the 10-inch brad-point bit to drill through the leg walls and the angled ends of the rails. Use the hammer to drive a 6½-inch peg through the hole until it's tight, being sure to keep the outer wall of the peg facing the wall of the leg. Use the Japanese knife or the fine-tooth saw to cut off the protruding peg ends. Repeat, pegging all the rails.

9 Now you'll cut and install the six 16-inch braces to give the table rigidity by putting tension on any loosely fitting pieces. Use the fine-tooth saw to make a fish-mouth cut at one end of each brace (see page 47), making the curve open enough to fit snugly against the leg. Repeat with all of the braces.

10 Repeat the pegging process outlined in step 8

to attach the braces to the legs. Use the brad-point bit to make a hole at an angle from the top of the brace down into the leg. Hammer in the 4-inch pegs, and cut the ends flush with the braces.

Making the Tabletop and Esterilla

1 On the 22-inch plywood square, mark a circle 20 inches in diameter. Do this by hammering a tack in the center of the plywood. Tie the string to the tack, then tie the marker to the string exactly 10 inches from the tack. Holding the string taut, mark a circle on the plywood. Use the jigsaw to cut out the circle, and smooth the edges with the belt sander.

2 Take the 20-inch bamboo lengths for the esterilla, and use the splitting knife and mallet to split each in half. Take five split lengths, and use the esterilla technique described on page 49 to flatten the bamboo. Use the

1"

Rail

← Tension brace pegging

4"

Figure 3

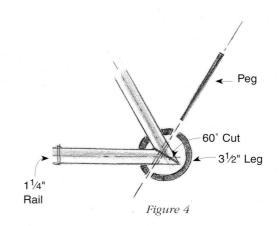

← Peg

60° Cut

3½" Leg

1¼" Rail

Figure 4

wood planer to shave off the rough inner wall to achieve a flat, uniform surface.

3 Refer to figure 5. Use the brush to apply a thin layer of glue to the plywood tabletop. Place the esterilla pieces on the glued surface. Start at one end of the tabletop and move across, adjusting the pieces so the nodes are lined up in the center.

4 Once all the pieces are laid out, lay a 22-inch wood strip across the tabletop center. Use the brad nails to temporarily secure the wood strips through the esterilla into the plywood, letting the nail heads stick up above

the wood just enough to be easily removed later with a hammer. As you move along, use your fingers to pull the edges of the pieces close together, making sure there are no gaps.

5 Cut the remaining wood strips into pieces, nailing the longer ones in the center of each half-circle and shorter ones along the edges. Let dry overnight.

6 With the wood strips still attached, turn the tabletop over, and use the fine-tooth saw to cut off the esterilla that over-hangs the edge. Pull off the wood strips and nails, and remove any glue with the wet rag.

7 Make the trim for the tabletop edge by splitting the 6-foot culm. From one half of the split culm, make a splint slightly wider than the thickness of the plywood and esterilla combined. Run the splint through the planer until it's flattened and thinned enough so it can be bent around the edge of the tabletop.

8 Now you'll install the trim; refer to figure 6. You'll need to work quickly; nearby in your work area, assemble the drill, the drill bit that's the same diameter as the furniture tacks, the hammer, and the tacks. Now, heat the strip. If you're using a propane torch, lay

Hold down strips

Nodes

Plywood
circle

Figure 5

Bamboo edge

Tack

Plywood
disk

Esterilla tabletop

Overlap

Bamboo edge

Figure 6

the strip on concrete outdoors, outer wall up, and run the flame over it in sweeping motions until it becomes loose and malleable. If you use a different flame source, hold the outer wall toward the flame.

9 Brush glue onto the inside of the trim and on the tabletop edge. Pick a starting point on the table edge. Drill a pilot hole through one end of the trim, press the trim to the edge, and hammer a tack through at the starting point. Continue attaching every 8 to 10 inches until you complete the edge. Overlap at the end by 3 inches, apply more

glue, drill another pilot hole, and hammer in the last tack. Use the fine-tooth saw to cut through the strip and slightly into the underlying trim to make a flat join. Use the wet rag to remove excess glue, and let dry.

Metric Equivalents

1/4"	6mm	2"	5.1 cm	14"	35.6 cm
3/8"	9.5 mm	3"	7.6 cm	16"	40.6 cm
1/2"	1.3 cm	3½"	8.9 cm	18"	45.7 cm
5/8"	1.6 cm	4"	10.2 cm	20"	50.8 cm
3/4"	1.9 cm	6½"	16.5 cm	21"	53.3 cm
1"	2.5 cm	8"	20.3 cm	22"	55.9 cm
1⅛"	2.8 cm	10"	25.4 cm	24"	61 cm
1¼"	3.2 cm	12"	30.5 cm	36"	0.9 m
				6'	1.8 m

Side Table
of Black Bamboo and Stone

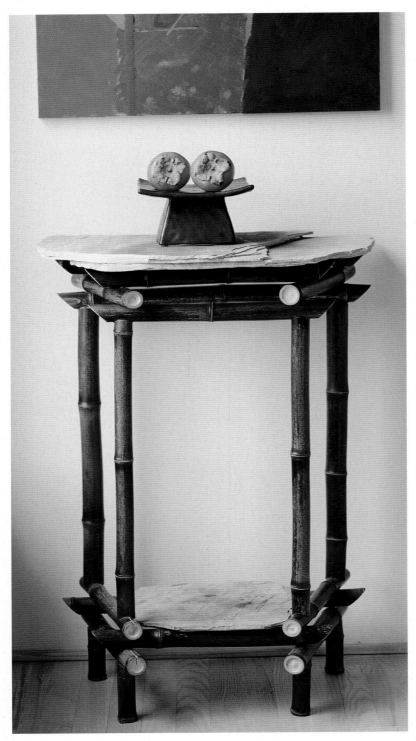

with assistance from Michel Spaan

THIS ELEGANT TABLE WAS DESIGNED TO COMPLEMENT THE STONE SLAB THAT SERVES AS ITS TOP. THE STAND IS PURPOSEFULLY SIMPLE, THE RICH BLACK OF THE BLACK BAMBOO (*PHYLLOSTACHYS NIGRA*) POLES ACCENTUATING THE RIPPLING LAYERS AND SUBTLE COLORS OF THE TENNESSEE SANDSTONE. FOR YOUR OWN SIDE TABLE, PICK ANY FLAT STONE WITH SIMILAR DIMENSIONS THAT PLEASES YOU. IF YOUR STONE NEEDS TRIMMING, YOUR LOCAL STONE YARD OR STONEMASON CAN DO IT FOR YOU.

Materials

Wood stain, brownish black color

4 wooden dowels, each 36 inches long, ⅜-inch diameter, to serve as inner core of legs

2 slabs of stone, relatively flat, approximate dimensions: one slab 23½ x 15 inches, ¾ to 1¼ inches thick, to serve as top; and one slab 12½ x 10½ inches, ⅜ to ½ inch thick, to serve as shelf

Tools and Supplies

Clean rag

Fine-tooth saw

Miter box

Drill press with ½-inch Forstner bit

Rebar, 48 inches long

Hammer

Measuring tape

Round rasp

Level

Toothpicks

Sanding sticks (see page 47), with medium and coarse sandpaper

Silicone glue

Cutting List

Description	Qty.	Material	Dimensions
Legs	4	Bamboo poles	1" x 21½"
Feet	4	Bamboo poles*	1" x 4½"
Front rails	3	Bamboo poles	1" x 16"
Back rails	3	Bamboo poles	1" x 22"
Side rails	6	Bamboo poles*	1" x 13"
Sleeves	4	Scrap bamboo	¾" x 36"

*With one end cut below the node

Instructions

1 Use the rag to apply two coats of the wood stain to the bamboo legs, feet, and rails. Let dry between coats.

2 Use the miter box and fine-tooth saw to cut the ends of the rails at 45° angles. Use the Forstner bit to drill a hole through both walls of the rails, 2½ inches from each end. Before you drill, make sure all the angled ends point in the same direction and that the two holes in each length are aligned since they will be slipped over the rigid dowels.

3 Use the rebar and hammer to knock out the diaphragms of the 21½-inch legs. Use the saw, rasp, and sanding poles to make fishmouth cuts (see page 47) at both ends of the legs.

4 Use the rebar and hammer to knock out the diaphragms of the 4½-inch feet, leaving the node at the bottom intact. Ream out the node membranes. Use the saw, rasp, and sanding sticks to make curved fishmouth cuts at the top of the feet.

5 Begin the stacking process by making sleeves for the feet. Insert the measuring tape into the

22" Side

3½"

16" Side

13" Side

2½"

2½" 11" 2½"

21½"

4"

hollow walls of each foot, and measure the distance from the base to the lower curved edge of the top. Transfer this measurement to a node-free length of the ¾-inch scrap bamboo and cut.

6 Place the end of a dowel in each foot. Insert the bamboo sleeve over the dowel into the foot, situating it between the dowel and the inner wall of the pole. Check for fit and make sure that the top edge of the sleeve is below the fishmouth cut. Repeat with the other three feet.

7 Slide a side rail onto two of the dowel assemblies. Use the level to determine if they're even. A slight unevenness may correct itself when other lengths are stacked, but if it is off more than ½ inch, correct it by cutting a new foot. If there are gaps where the curved top of the feet meet the horizontals, rework the fishmouth cut to get a smooth joint. Repeat with the other side rails and two feet.

8 Now remove the side rails, dowel, and sleeve and squeeze a generous amount of silicone glue into the base of the foot. Reinsert the dowel and the sleeve into the foot. At the top, apply more glue, using a toothpick to work the glue into any spaces. The glue will help keep the stand from wobbling. Repeat with the other side.

9 Join the sides together by sliding a back rail and a front rail over the sets of dowels. The structure is now beginning to look like a four-sided table base! Use the level and adjust. Now, slide a pair of side rails through the dowels.

10 Slide the legs over the dowels, checking for a good fit at the fishmouth joint. Then slide the front and back rails onto the dowels, again checking that the fishmouth cut fits well against the bamboo. When you're satisfied that the pieces fit together and are level, remove the front and back horizontals and the legs. Repeat the technique in step 5 to make sleeves for the top and bottom ends of each leg. (The sleeves for the legs don't have to fill the entire length of the leg; they can be just long enough to stabilize the assembly.) Be sure to squeeze glue into the sleeve and dowel assemblies and use a toothpick to push it in.

11 Restack the last of the rails, checking and adjusting each one for level and fit. At this point, the stand is now assembled. Saw the dowels 1 inch above the top rails. Center the stone slab on the top of the dowels. Check with the level; if all is well, saw off the dowels flush with the bamboo, then rest the slab directly on the bamboo. If the slab is irregular, mark and cut each dowel so the stone rests solidly.

12 Place the smaller stone on the lower horizontals to create the shelf.

36" Dowel

1" Leg

Diaphragm removed

Dowel Assembly Detail

Dowel

Sleeve

Dowel

Sleeve

Metric Equivalents

⅜"	9.5 mm
½"	1.3 cm
¾"	1.9 cm
1"	2.5 cm
1¼"	3.2 cm
2½"	6.4 cm
4½"	11.4 cm
10½"	26.7 cm
12½"	31.8 cm
13"	33 cm
15"	38.1 cm
16"	40.6 cm
21½"	54.6 cm
22"	55.9 cm
23½"	59.7 cm
36"	91.4 cm
48"	121.9 cm

Elegant Vases

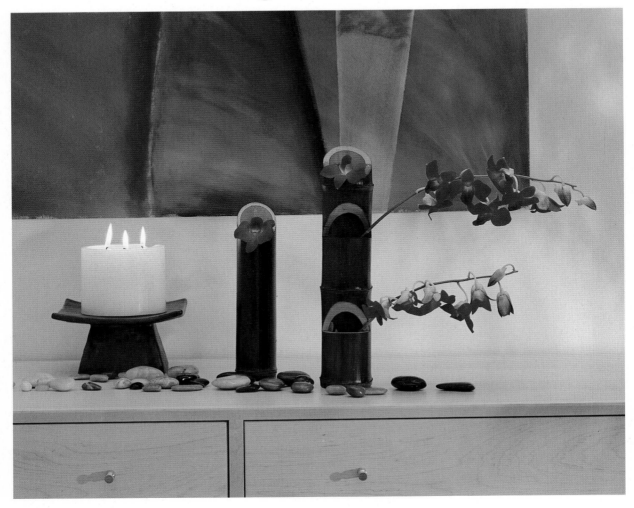

ITH A FEW WELL-PLACED CUTS, YOU CAN TRANSFORM CAST-OFF LENGTHS OF BAMBOO INTO SOPHISTICATED FLOWER VASES. LENGTHS TAKEN FROM THE UPPER MID-CULM MAKE LONG, SLENDER VASES, WHILE THOSE TAKEN FROM THE BASE OF THE CULM PROVIDE CLOSE-SPACED NODES FOR MULTIPLE FLOWER OPENINGS. THE INTERNODES HOLD SMALL AMOUNTS OF WATER, ENOUGH TO KEEP CUT FLOWERS FRESH FOR SEVERAL DAYS. IF YOU WISH TO HAVE A HANGING VASE, DRILL A HOLE AT AN UPWARD ANGLE BELOW THE TOP NODE IN THE BACK. HAMMER A FINISHING NAIL IN THE WALL AND HANG THE VASE.

Instructions

Simple Vase

1 Working with the 11-inch length of bamboo, make a cut straight across the base ½ inch below the node, so the vase sits upright.

2 Select and mark the side of the culm to be the front of the vase. Do this by rotating the piece, looking for interesting scarring or coloration. Place the bamboo in the miter box, and line up the top with the 45° angle slots in such a way that the elongated cut you are about to make will open at the front. Saw, being careful not to feather the outer wall.

3 Use the sanding sponge to lightly sand the edges. Brush the wood stain on the outer walls. Let dry. If desired, use the sealer to coat the inside of the culm.

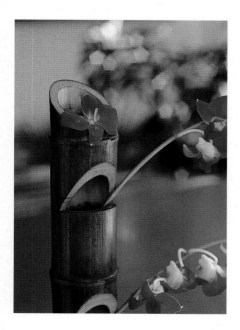

Vase with Two Openings

1 At the base of the 14-inch length of bamboo, use the saw to make a cut straight across, ½ inch below the node, so the vase sits upright.

2 Choose the side of the culm to serve as the front of the vase. Place it in the miter box and line up the top of the bamboo with the 45°-angle slots in such a way that the elongated cut you are about to make will open toward the front. Use the saw to make a diagonal cut ¼ inch above the front of the node and 2 inches above the back of the node.

3 Now you'll mark the placement of the holes. Measure halfway up from the bottom node and mark a line across. Mark the same distance above the middle node. Cut straight across each line, ⅝-inch deep into the pole.

4 Mark 1½ inches above the bottom cut, and secure the bamboo firmly in place. Position the saw at the mark, angling it into the culm and down, along an imaginary line that will meet the inside ends of the cut. Saw, keeping the blade at an steady angle, until you reach the cut. Pop out the piece. Repeat to create the upper hole, this time starting the cut 2 inches above the cut.

Materials

1 length of black bamboo, 11 inches long, 2⅜ inches in diameter, with one node at the base

1 length of black bamboo 14 inches long, 2¾ inches in diameter, with three nodes

Wood stain, brownish black color

Tools and Supplies

Fine-tooth saw

Pencil

Miter box

Sanding sponge

Paintbrush

Wood sealer (optional)

Metric Equivalents

¼"	6 mm
½"	1.3 cm
⅝"	1.6 cm
1½"	3.8 cm
2"	5.1 cm
2⅜"	6 cm
2¾"	7 cm
11"	27.9 cm
14"	35.6 cm

5 Use the sanding sponge to lightly sand the edges of the vase, and brush the stain on the outer walls. Let dry.

6 If desired, coat the inner walls with the sealer and let dry.

Triple Serving Vessel

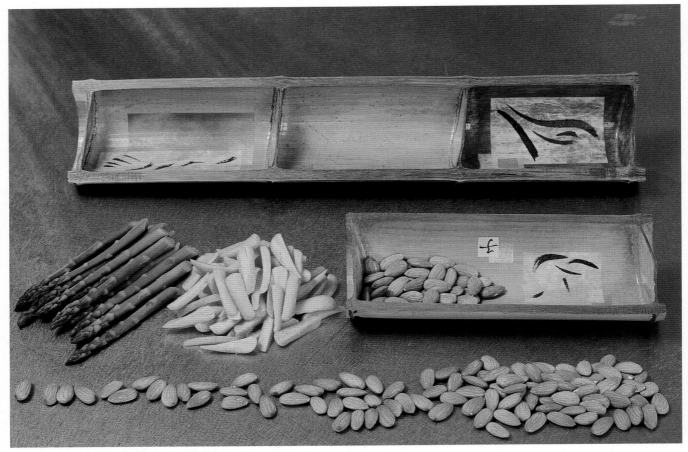

Designer, Peter Gallagher

IANT *MOSO* BAMBOO FROM JAPAN COMBINES WITH EARTH-TONED WOOD STAIN AND SILVER LEAF TO CREATE THIS SERVING VESSEL WITH THREE SPACIOUS COMPARTMENTS. FINE ASIAN PAPER IS INFUSED WITH THE BLACK INK OF *SUMI* BRUSH STROKES AND LAID WITHIN THE CURVED HOLLOWS OF THE BAMBOO. YOU'LL WANT TO PRESENT VEGETABLES, FRUITS, OR NUTS IN THIS CONTAINER. IF IT NEEDS CLEANING, HAND WASH WITH A MILD DETERGENT, AND WIPE GENTLY.

Instructions

1 Use the scrub brush and bleach solution to wash the outside of the bamboo. Dry with the cloth.

2 Use the saw to cut straight across the ends of the bamboo, close to the nodes.

3 Use the splitting knife and mallet to split the bamboo. Smooth the edges with the knife or shaver, then sand until smooth.

4 Use the scrub brush and bleach solution to clean the diaphragms and inner walls thoroughly. Let dry.

5 Use the artist's brush to apply the brownish black stain to the edges of the container, and the plum stain to one of the inner compartments. Let dry.

6 Use the cloth measuring tape to measure the inner dimensions of the unpainted inner sections of the vessel. Use the ruler and marker to transfer the dimensions to the white paper, and use the scissors to cut out the patterns. Outline the patterns on the Japanese paper, then cut out.

7 Thin the white glue with water, then brush it onto the back of the pieces of Japanese paper. Adhere the papers to the inner walls. Let dry.

8 Use the brush to apply four coats of the varnish to seal the vessel, allowing it to dry between coats.

Materials

1 length of moso bamboo (*Phyllostachys pubescens*) 24 inches long, 4½ inches in diameter, with four nodes taken from the base of the culm

Wood stains, brownish black and plum colors

Handmade Japanese paper decorated with sumi brushwork and silver leaf

Food-safe varnish

Tools and Supplies

Scrub brush

Mild water and bleach solution

Cloth

Fine-tooth saw

Splitting knife

Mallet

Knife or shaver

Sanding sponge

Artist's brush

Cloth measuring tape

Ruler

Pencil

Scissors

White glue

Brush for applying glue

Small paintbrush

Metric Equivalents

| 4½" | 11.4 cm |
| 24" | 61 cm |

Bamboo vessels.
DESIGN AND PHOTO BY PETER GALLAGHER

Sushi Trays

 OUR SUSHI MAY COME FROM THE TAKE-OUT COUNTER OF A LOCAL

JAPANESE RESTAURANT, BUT YOU CAN SERVE IT ON YOUR OWN

HANDCRAFTED TRAYS OF FRESH GREEN BAMBOO. THE TRAYS ARE

A GOOD WAY TO PRACTICE SPLITTING, TOO.

Materials

3 fresh, green lengths of bamboo, each 6½ inches long, 3 inches in diameter, without nodes

Metric Equivalents

3"	7.6 cm
6½"	16.5 cm

Tools and Supplies

Water

Bleach

Scrub brush

Cloth

Measuring tape

Fine-tip permanent marker

Fine-tooth saw

Splitting knife

Mallet

Instructions

1 Use the scrub brush and a weak solution of water and bleach to wash the outside of the bamboo. Dry with the cloth.

2 Use the hammer and mallet to split each length into four pieces, making a total of 12 splints.

3 Use the knife or shaver to lightly remove the rough edges along the sides of the trays. You're ready to serve!

Tea Cups and Napkin Rings

Instructions

Making the Teacups

1 Use the scrub brush and water to clean the outside of the 4-inch pieces of bamboo. Wipe dry with the cloth.

2 Use the saw to cut each length evenly across the base, ⅜ inch below the node, so the piece sits upright.

3 Measure and mark 3½ inches up from the base of each piece.

4 Cut at the mark you made, sawing the length carefully to prevent shredding. Sand the sharp edges of the inner and outer rim until smooth.

5 Use the bottle brush and water to scrub the inside of the cups, then turn them upside down to dry.

S IMPLICITY IS THE KEY TO ARTFUL ENTERTAINING, AND THESE BAMBOO DINING UTENSILS ARE LOVELY TO LOOK AT AND USE. THEY'RE SO EASY TO MAKE, PERHAPS YOU'LL WANT TO MAKE SOME AS TAKE-HOME GIFTS.

Materials

4 fresh, green lengths of bamboo, each 4 inches long, 1¾ inches in diameter

1 fresh, green length of bamboo, 8 inches long, 1¾ inch in diameter

Tools and Supplies

Scrub brush

Cloth

Fine-tooth saw

Measuring tape

Fine-tip permanent marker

Knife

Sanding sponge

Bottle brush

Metric Equivalents

⅜"	9.5 mm
1⅝"	4.1 cm
1¾"	4.4 cm
3½"	8.9 cm
4"	10.2 cm
8"	20.3 cm

Making the Napkin Rings

1 With the scrub brush and water, wash the outside of the 8-inch length of bamboo. Dry with the cloth.

2 Measure and cut four pieces from the 8-inch length, each 1⅝ inches long. Use the sanding sponge to smooth the ends.

Shoji Window Screen

with assistance from Randall Ray and Jean Clark

THIS WINDOW TREATMENT WAS INSPIRED BY JAPANESE *SHOJI* SCREENS, SLIDING WALL

PANELS OF HEAVY PAPER FRAMED WITH WOOD. THE SPLINTS ARE CRAFTED FROM ONE

LENGTH OF BLACK BAMBOO, ADDING TO THE SYMMETRY OF THE SQUARE AND CIRCLE

SHAPES, WHILE THE PROMINENT, IRREGULAR NODES OF THE BAMBOO ADD TEXTURAL CONTRAST.

THE *KOZO* PAPER, SHOWN IN THE WINDOW SCREEN IN THE PHOTOGRAPH, IS MADE FROM MULBERRY

FIBER AND IS SIMILAR TO TRADITIONAL SHOJI PAPER.

Instructions

1 To determine where to put the center of the circle on the plywood square, mark the midpoint of each side at 10 inches, then draw lines that connect the opposite midpoints, intersecting in the middle. This is the center of the square. Drive a nail into the center. Tie the string to the nail, then tie the string to the pencil, exactly 8 ½ inches from the nail. Keeping the string taut, mark the circle on the plywood.

2 Drill a ⅝-inch hole, or one large enough to accommodate the jigsaw blade, inside the circle near the circle's edge. Saw from the hole to the outline, approaching at an angle, then saw around the circle. Remove the circular piece of scrap. Use the sanding sponge to smooth the inner edges.

3 Brush the plywood frame and lengths of molding with two coats of the wood stain. Let dry between coats. Apply the sealer and let dry.

4 Use the saw and miter box to cut four lengths from the molding, each 21 inches long, with the ends mitered to a 45° angle so the pieces of molding fit together in a square. Use the finishing nails to attach the molding to the edge of the plywood, creating the outer frame.

Materials

Scrap plywood, ½" x 20" x 20"

20 1¼-inch ringed nails

1 sheet of Japanese shoji paper, 20" x 20"

Wood stain, walnut color

Wood sealer

20 1-inch finishing nails

Double-sided tape

2 ½-inch screw eyes

2 cup hooks

Monofilament

Tools and Supplies

Measuring tape

Pencil

Hammer

15-inch piece of string

Power drill and assorted drill bits

Jigsaw

Sanding sponge

Miter box

Wood glue

Paintbrush

Nail punch

Cutting List

Description	Qty.	Material	Dimensions
Verticals	3	Black bamboo poles	¾" x 17¼"
Horizontals	1	Black bamboo pole, curved, split in half	¾" x 17¼"
Top, bottom, and center horizontals	3	Bamboo splints	1¾" x 16"
Frame base	1	Plywood	½" x 20" x 20"
Outer frame	1	Molding with rounded edges	½" x 1½" x 48"

Metric Equivalents

½"	1.3 cm	2"	5.1 cm	15"	38.1 cm
⅝"	1.6 cm	2½"	1.3 cm	15½"	39.4 cm
¾"	1.9 cm	5"	12.7 cm	16"	40.6 cm
1"	2.5 cm	8"	20.3 cm	17¼"	43.8 cm
1¼"	3.2 cm	8½"	21.6 cm	20"	50.8 cm
1½"	3.8 cm	10"	25.4 cm	21"	53.3 cm
1¾"	4.4 cm	12½"	31.8 cm	48"	1.2 m

2" 2"

17"
Cutout

½"
Molding

15½"

12½"

10"

8"

5"

20"

5 On a flat work surface, place the frame on top of the 20 × 20-inch piece of plywood. Cut and mark one of the three vertical poles so it will fit directly over the center of the frame. Cut and mark the other two poles at a slight angle, so they can be fit snugly on the frame 2 inches on either side of the center pole. Turn over the frame so you're now working on the back of the frame. Attach the vertical poles by drilling a pilot hole through the ends of the bamboo at an angle into the wood. Hammer the ringed nails into the plywood, and use the nail punch to drive the nail heads further in. Turn the frame over to the front.

6 Lay the horizontal pieces across the verticals. From bottom to top, measure and mark the center points at 5, 8, 10, 12½ and 15½ inches. Mark and cut the ends of the center horizontal, and cut the other ends at an angle to follow the curve.

7 Attach the middle piece to the left and right bamboo verticals. Drill a pilot hole using a bit the same size as the ringed nail. Coat a ringed nail with wood glue. Insert the nail through the bamboo wall, and gently tap the head with the nail punch and hammer.

8 Attach the other bamboo pieces. Drill pilot holes in the corners, and hammer in ringed nails, followed by a tap with the hammer and nail punch. No need to add glue, as the nails firmly grab the plywood.

9 ffix the double-sided tape to the back of the frame, ¾ inch from the inside of the molding. Carefully center the paper and lay the edges over the tape, pressing down along all four sides to secure.

10 Attach the screw eyes to the top of the frame 2 inches from each side. Tie the monofilament to the screw eyes, and hang from the cup hooks at the desired location.

Bamboo Doormat

Designers, Yucatan Bamboo

THIS HANDSOME MAT OF IRON BAMBOO (*DENDROCALAMUS STRICTUS*) WORKS WELL AT THE EDGE OF A PATIO, PORCH, OR POOL. ORIGINALLY DESIGNED FOR THE STONE COURTYARDS OF A MEXICAN HACIENDA-TURNED-HOTEL, THIS RESILIENT, TIGHTLY STRUNG MAT KEEPS FEET ABOVE GROUND LEVEL AND ALLOWS WATER TO DRAIN THROUGH. A DRILL PRESS IS NEEDED FOR THIS PROJECT; IF YOU DON'T HAVE ONE, TAKE THE PIECES TO A LOCAL WOOD SHOP. THE PROCESS FOR MAKING THIS DOORMAT IS FULLY PATENTED, BUT INDIVIDUALS ARE WELCOME TO MAKE MATS FOR THEIR PERSONAL USE.

Instructions

1 Set up the jig on the drill press. Place the 36-inch plank on the surface of the press, up against a straight edge. Make sure the plank is level. Lower the drill press with the ⅜-inch bit until it makes an indentation in the plank. Measure 3 inches beyond this point, and attach with screws the 6-inch strip of wood to serve as the stop. Make sure the stop is square at a 90° angle to the plank. Line the lengths of bamboo up against the stop to ensure all drilled holes will be a uniform 3 inches from the end.

2 Examine each pole to determine where the hole should be drilled. If the pole has a slight curve, rotate it so the inside of the curve faces upward. Place one end up against the stop with the section to be drilled lying flat against the plank under the drill press. It's acceptable for part of the pole to curve slightly upward off the plank. Use the drill press to make the hole, repeating the process (at one end only) with all of the lengths.

3 Measure 27 inches from the stop, exactly in line with the drill press indentation, and mark. Drill a ⅛-inch pilot hole at the mark and through the plank at a 90° angle. Turn over the plank and drive the 3-inch nail through it. Return the plank to the drill press, the tip of the nail pointed up.

Materials

1 x 6 wood plank, 36" long

1 strip of wood, ½" x 1" x 6"

5 drywall screws, 1⅝ inches long

3-inch nail, ⅜-inch diameter

2 lengths of ¼-inch galvanized
steel rod, each 24 inches long

String

Wood sealer

Tools and Supplies

T-square

Drill press with ⅜-inch drill bit

Level

Measuring tape or ruler

Hammer

Vise

Pliers

Hacksaw

Splitting knife

Mallet

Power drill with assorted drill bits

Paintbrush

Cutting List

Description	Qty.	Material	Dimensions
Mat surface	20	Iron bamboo pole	¾" to 1" x 30"
Feet	1	Iron bamboo pole*	2" to 2½" x 22"

*It's important to select a very straight length of bamboo.

Metric Equivalents

⅛"	3 mm	3"	7.6 cm
¼"	6 mm	6"	15.2 cm
⅜"	9.5 mm	22"	55.9 cm
½"	1.3 cm	24"	61 cm
¾"	1.9 cm	27"	68.6 cm
1"	2.5 cm	30"	81.3 cm
1⅝"	4.1 cm	36"	0.9 m
2"	5.1 cm		
2½"	6.4 cm		

4 Slip the drilled end of a pole over the nail, pressing the other end against the stop. Drill a hole through the bamboo. Repeat with all the poles.

5 Use the vise and pliers to make a ½-inch, 90° bend at the end of each steel rod.

6 Working with one pole at a time, determine its curvature for placement on the rods. String the two rods through the holes in the bamboo, alternating the slight curve facing up with the slight curve facing down, to create a flat surface. Use the mallet to hammer each length close to the next. String all the poles.

7 Use the pliers and hammer to bend the wire rods at a right angle, securing the poles together. Use a hacksaw to cut off the excess.

8 Now you'll attach the feet. Use the knife and mallet to split the 22-inch length. Line up the halves along the wire rods, with the curved side up against the bamboo and the flat side resting evenly on the floor.

9 Use the power drill and screws to attach the feet to the bottom of the mat, driving them in from the under side of the mat.

10 Brush the wood sealer on all surfaces of the mat and let dry.

Chopsticks & Chopsticks Rests

OU CAN PRACTICE SPLITTING BAMBOO BY CRAFTING YOUR OWN CHOPSTICKS AND

CLEVER RESTS TO SET THEM ON. TIE TOGETHER A BUNDLE WITH RIBBON OR RAFFIA,

AND YOU HAVE A CHARMING GIFT THE NEXT TIME YOU DINE AT A FRIEND'S HOUSE.

Instructions

Making the Chopsticks

1. Use the scrub brush and a weak solution of water and bleach to wash the outside of the bamboo. Dry with the cloth.

2. Use the knife and mallet to split the length in half. Reserve half to make the chopstick rests, and split the other half into quarters.

3. Take one of the four splints you just made. Use the knife or shaver and sanding sponge to smooth the rough edges and the inside wall until the bamboo feels good in your hand.

4. Use the knife to shave off the outer corners and form a blunt-ended point on one end of the chopsticks. This end will serve to pick up food. Sand the edges.

5. Hold the splint upright with its node end down. Place the splitting knife exactly in the center of the blunt-pointed end. Tap the knife with the mallet. Carefully twist and move the knife to split to, but not through, the node. The person who uses the chopsticks will pull them apart.

6. Repeat steps 3 through 5 with the remaining three splints.

Chopstick Rests

1. Take the split half you put aside in step 2 above. Use the saw to cut it into four pieces, each ¾ inch wide. Lay the resulting half-circle pieces of bamboo on their sides.

2. With the splitting knife and mallet, remove ½ inch from the ends of the half circles, so that when you turn them upright to hold the chopsticks, their height is ¾ inch.

Materials

1 fresh, green length of bamboo, 10¼ inches long, 2¼ inches in diameter, with one node 1½ inches from the base

Tools and Supplies

Water
Bleach
Scrub brush
Cloth
Splitting knife
Mallet
Knife or shaver
Sanding sponge

Metric Equivalents

½"	1.3 cm
¾"	1.9 cm
1½"	3.8 cm
2¼"	5.7 cm
10¼"	26 cm

Formal Japanese Flower Container

Designer, Brother Konomo Utsumi

T HIS ELEGANT CONTAINER FOCUSES ON THE SIMPLICITY, FORM, AND UTILITY OF BAM-
BOO. WHEN YOU MAKE THIS FLOWER CONTAINER, YOU'LL BE CREATING AN OBJECT
THAT REFLECTS CONTEMPORARY JAPANESE AESTHETICS.

Instructions

1 Sweat the bamboo poles, using the process explained on page 50.

2 Cut a 7-inch length from one of the poles. Use the knife and mallet to split it, creating two feet.

3 Arrange the three poles side by side. Position the center area of each length, where the openings will be cut, between nodes. Mark the center points.

4 Keeping the centers lined up and measuring from the left side, mark one pole at 35 inches, one at 21 inches, and one at 27 inches. Use the saw to cut the lengths straight across at the nodes. Cut the other ends on an angle to create an open elongated oval.

5 Now you'll make the center openings. Starting with the 35-inch length pole, measure 14 inches from the left end, and use the saw to cut straight down about halfway through the culm. From that cut, measure up 1¾ inches and saw on an angle into the culm to meet the cut. Pop out the inner piece. Repeat this process with the other two poles, cutting the 21-inch length 9 inches up from the base, and the 27-inch piece 12 inches up from the base.

6 Use the knife to slice off the black outer layer of the bamboo and to make decorative stripes on the ends, at the center openings, and on the feet.

7 Attach the feet by centering the three poles with their openings facedown. Measure and mark 5 inches below the straight cut of the opening and 7 inches above it.

8 Center the feet, curved sides facing down, along the points you marked. Drill pilot holes, and drive the 1-inch screws through the feet and into the poles.

9 Use the brush to apply the sealer on the outer surfaces and inside the center and end openings of the vase. Let dry.

Materials

3 poles of black bamboo (*Phyllostachys nigra*) each 36 inches long, 1 inch in diameter

Metric Equivalents

1"	2.5 cm
1½"	3.8 cm
1¾"	4.4 cm
5"	12.7 cm
7"	17.8 cm
9"	22.9 cm
12"	30.5 cm
14"	35.6 cm
21"	53.34 cm
21¾"	55.24 cm
27"	68.6 cm

Tools and Supplies

Fine-tooth saw

Splitting knife

Mallet

Fine-tip permanent marker

Measuring tape

Knife

Power drill and assorted drill bits

6 drywall screws, 1 to 1½ inches long

Water sealer

Paintbrush

Chapter 8

Fences and Screens

YOU CAN USE BAMBOO TO BUILD FENCES AND SCREENS IN A WIDE RANGE OF STYLES TO COMPLEMENT GARDENS, PONDS, COURTYARDS, DECKS, AND PATIOS. A LOW, SEE-THROUGH FENCE CAN EASE THE TRANSITION BETWEEN DIVERSE AREAS IN A LANDSCAPE, WHILE A TALL, NARROW SCREEN ADJACENT TO A HOUSE CAN SUGGEST A

SENSE OF PRIVACY. SOLID WALLS OF BAMBOO CAN CONVENIENTLY SHUT OUT AN UNDESIRABLE VIEW, WHILE AN OPEN BAMBOO FENCE CAN EMBRACE THE LANDSCAPE BEYOND. ONCE YOU UNDERSTAND THE BASIC CONSTRUCTION PROCESSES FOR MAKING FENCES AND SCREENS, YOU'LL BE ABLE TO CREATE VARIATIONS THAT SUIT YOUR LANDSCAPE AND REFLECT YOUR PERSONAL STYLE.

Fence and Screen Construction

Fences and screens consist of five main elements as shown on page 102. *Posts* are placed upright in the ground, joined together by *stringers*. *Verticals* are bamboo poles secured onto the stringers, and *horizontals* are decorative split lengths that attach to the verticals and stringers. *Ties* lash over the horizontals to secure them to the poles and stringers.

DESIGN AND LOCATION

Planning is the key to a successful fence or screen. Measure the intended site, and use stakes and flagging to mark its boundaries. Draw the design on graph paper, drawing to scale the length and height of the structure and the placement of stringers and verticals. Based on your own plan or any modifications you make in the following projects, develop a list of materials and supplies. Where do you build? Multi-paneled fences, as described in this book, are built on-site. Single panel screens can be constructed off-site and then installed.

POSTS

Spaced 5 to 8 feet (1.5 to 2.4 m) apart, posts act as anchors for the entire structure. Fence posts and timbers are available from home improvement and farm supply stores; buy posts that have been pressure-treated with chemicals to withstand the weather and insects. If the post is untreated, brush a wood preservative over the part of the post to be sunk in the ground. If you plan to use bamboo posts, choose stout, strong poles. When calculating length, add a minimum of 18 inches (45.7 cm) to set in the ground.

NOTCHES

Measure and mark the center placement of the stringers on the post, i.e., the point at which the middle of each stringer will join the post. From that point, draw parallel lines to mark the width of the stringer. Saw along these lines to a depth of ¾ inch (1.9 cm) or more, then use a hammer and chisel to knock out the wood in between. At this point, stain the posts if desired and let dry.

SETTING POSTS

Using a post hole digger, dig holes to a depth of 18 to 24 inches (45.7 to 61 cm). Be mindful of the possibility of underground electrical and television cables, irrigation hoses, and water and gas lines. Place 2 inches (5 cm) of crushed rock into the hole, and tamp. Place the posts in the holes and fill in

Split bamboo creates diamond patterns over panels of reed fencing.
DESIGN AND PHOTO BY ROBERT SMALL

Bamboo poles are attached on the diagonal to create this fence at the Bamboo Farm, Savannah, Georgia.

In this bamboo fence, "windows" open to the landscape beyond.
DESIGN AND PHOTO BY ROBERT SMALL

Figure 1

Labels in figure:
- Guide line
- Horizontal stringer
- Landscape timber or round post
- Guide line
- Vertical poles
- 6" Soil on top
- Crushed rocks around post
- 6" Crushed rocks on bottom

6 inches (15.2 cm) with the crushed rock. Use a piece of rebar to jiggle and tamp the rocks to remove air pockets to stabilize the pole. Use a level to check that the post stands straight, and tamp the rock again. Pour concrete or add gravel, and check the post again with the level. Continue to add gravel or concrete. Just below ground level, fill with soil.

STRINGERS

Pressure-treated 2 x 2 lumber makes good, all-purpose stringers for bamboo fences and screens. It's relatively light, easy to work with, and gives a flat surface on which to attach the verticals. Bamboo can also be used as stringers, but be sure to choose straight, even poles. After measuring and cutting the stringers, stain them (if desired) and let dry. To attach stringers to posts, position them and screw into place.

VERTICALS

Filling in the framework with vertical bamboo poles is the fun part! Verticals should not rest directly on the ground or decay will set in. It's best to secure verticals several inches above ground level or rest them on a stone wall. Likewise, for outdoor projects, it's not a good idea to let rain collect in bamboo's upright hollow tubes. For this reason, when you're cutting vertical poles, plan so you cut just above the top node. That way what little water collects in the concave node membrane will evaporate quickly.

You can install a guide string to keep the top and bottom of the vertical bamboo poles in line.

Mark the posts with the position, and drill a 1-inch (2.5 cm) screw into each mark. Stretch mason's cord from screw to screw. Use this as well as the T-square to keep the poles even.

When working with long or heavy vertical poles, you can add a temporary ledge of 1 x 4 lumber for the poles to rest upon as you attach them to the stringers. Drive in screws at an angle through the ends of the board between the posts. After you've attached the verticals, remove the screws and ledge.

ATTACHING VERTICAL POLES

Work on one panel of your fence or screen at a time. Position a pole next to one post, and attach it by drilling a pilot hole through it and the stringer, and then nailing or screwing it into place. For the next vertical, try out different poles, rotating, and reversing tops and bottoms to find the best fit. When you find the pole and placement that feels right, clamp or hold in place, drill pilot holes, and use screws or nails to attach. Work back and forth from post to post, attaching verticals in groups until the poles meet in the middle. Choose the last group carefully to make sure there are no major gaps.

Drilling at an angle

Landscape timber or round post

Figure 2

HORIZONTALS

Made of split bamboo, horizontals visually unify fences or screens, and hide screw or nail heads. To make a horizontal, choose a straight, handsome pole 2 to 3 inches (5.1 to 7.6 cm) in diameter. Carefully split it in half, to give you two lengths. Position the split bamboo horizontals on top of the verticals, which are positioned in turn over the stringers. Measure, mark, and cut the ends of the horizontals so they fit close against the post. Work with an assistant to secure them temporarily with twine to the verticals and stringers. Once they're positioned to your satisfaction, screw or tie into place.

TIES

Decorative Japanese ties are used minimally, as accents. Three to five ties per horizontal is typical, with placement often staggered. For your design, try out different arrangements on paper, then transfer the positions onto the horizontals with removable tape. Stand back and see if the pattern is too busy or too sparse, adjusting until you get a pleasing arrangement. Avoid an ambitious array until you feel comfortable making the knots. To prevent the twine from roughening and coloring your fingers, wear latex or tight-fitting work gloves.

If the poles are close together, ties are difficult to make since one pair of arms can't reach over and around the poles. It's a good idea to enlist the aid of a friend who can stand behind the fence, pull the cordage through to the back, manipulate it, and pass it back to you. If you have difficulty getting the twine through the gaps between the poles, wrap the ends with masking tape. This forms a needle-like cap which you will cut and discard when the tying is complete.

Tools and Supplies

To lay out the area and boundaries: measuring tape, stakes, flagging, mason's cord

To cut posts and stringers: power saw

To make notches in the posts: hand saw, hammer, and chisel

To dig and set posts: shovel, post-hole digger, tamper, 4-foot (1.2 m) length of rebar, crushed rock or gravel, quick-set concrete, water, and bucket or other container in which to mix the concrete

To keep fence components level: level, T-square

To attach stringers to posts and vertical poles to stringers: quick-release clamps, power drill, assorted drill bits, nails, and screws

To color and/or seal the posts: stain, paint, or sealer, paintbrush

Kenninji Fence

with assistance from Allen Fowler

HANDSOME IN ALL ITS VARIATIONS, THE *KENNINJI-GAKI* IS THE MOST COMMON SCREENING FENCE IN JAPAN. MADE WITH WHOLE OR SPLIT BAMBOO, IT HAS THREE TO FIVE SPLIT HORIZONTALS AND USES DARKENED POSTS AND TIES FOR CONTRAST. THE CLOSE PLACEMENT OF ITS POLES CREATES AN ATTRACTIVE PRIVACY WALL AS WELL. THE RIGHT-ANGLED, THREE-PANELLED FENCE SHOWN HERE HIDES TWO LARGE AIR-CONDITIONING UNITS. HARVESTED FROM THE OWNER'S ADJACENT BAMBOO GROVE, THE POLES COMPLEMENT THE STONE WALKWAY AND EARTHEN WALLS, AND PROVIDE A WARM AND SOOTHING ENTRANCE.

Materials

Wood stain, deep brown

16   2¼-inch galvanized decking screws

136   1½-inch galvanized decking screws

12   3-inch galvanized decking screws

Double-strand, black hemp twine

Tools and Supplies

Fence-building tools and supplies listed on page 103

Fine-tooth saw

T-square

Scissors

Binder twine

Removable tape

Instructions

1 You can adapt these instructions to make as many panels as you require. First, mark the posts where the stringers will be attached. Measuring from the top of each post, mark at 2, 16, 28, and 42 inches. Make two more marks, one at 54 inches, to mark the bottom of the vertical poles, and one at 58 inches, to mark the ground line. The remaining 18 inches of the post will be sunk into the ground.

Metric Equivalents

1"	2.5 cm	28"	71.1 cm
1½"	3.8 cm	42"	106.7 cm
2"	5.1 cm	54"	137 cm
2¼"	5.7 cm	57"	145 cm
2½"	6.4 cm	58"	147.4 cm
3"	7.6 cm	60"	152.4 cm
3½"	8.9 cm	63"	160 cm
16"	40.6 cm	64"	163 cm
18"	45.7 cm	76"	100.6 cm

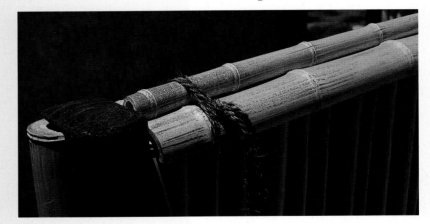

2 Brush the posts with the dark brown stain and let dry.

3 Measure and mark notches in the posts to accommodate the stringers. Saw to a 1-inch depth at the notch lines, and use the hammer and chisel to remove the wood in between.

Cutting List

Note: Quantities and instructions are given for 1 fence panel 64 inches long and 57 inches high.

Description	Qty.	Material	Dimensions
Posts	2	Pressure-treated landscape timbers	2½" x 3½" x 76"
Stringers	4	Pressure-treated 2 x 2 lumber	63" long
Verticals	36	Bamboo poles	1½" x 54", one end cut just above the node
Horizontals	5	Split bamboo lengths and cap	2" x 60"
Top of cap	1	Bamboo pole	1½" x 60"

4 Dig holes for the posts, and set the posts in place with gravel.

5 Place the stringers into the notches on the posts. Secure with the 2¼-inch screws.

6 Attach the vertical poles to the stringers with the 1½-inch screws. Use the T-square to keep the poles straight. Beware of placing the poles too close together; you'll need gaps wide enough for several strands of twine to pass through when you make the ties.

7 Use the binder twine to temporarily attach the split horizontals over the poles to cover the screw heads. Use the 3-inch screws to secure to the stringer, and remove the twine.

8 Now add the caps to the panel. Rest one split length, open side down, on the tops of the poles. Place the other split length beside it, resting on the stringer. The caps should slightly overlap the edges. Drill pilot holes, then use the 3-inch screws to secure the split bamboo horizontals, poles, and stringers. Position the whole pole in the crevice between the split lengths. Using the 3-inch screws, secure to the top stringer.

9 Use the double-strand black twine to tie traditional Japanese ties (see page 45), covering the screw heads on the split horizontals. Along the cap, use more twine and make a double overhand knot at the ends.

Caps ½" top

2 x 2

Horizontal 2" cap

2 x 2

2 x 2 Stringer

Cap

Top

Post

Corner pole

Post

1"

16"

28"

42"

1"

58"

54"

76"

4"

Teppo Screens on Stone Ledges

Designer, Michel Spaan

THESE MAGNIFI-
CENT SCREENS
CREATE AN
ENCLOSURE THAT SERENELY
EMBRACES THE QUIETLY FLOW-
ING WATER OF THE FOUNTAIN.
THE LEFT AND RIGHT SCREENS
DIFFER IN HEIGHT BECAUSE OF
THE SLOPING TERRAIN, BUT
THEIR TOPS ARE LEVEL WITH
EACH OTHER. (THESE
INSTRUCTIONS ARE ADJUSTED
TO MAKE SCREENS OF EQUAL
HEIGHT.) ATTACHMENTS TO
UPPER TREE BRANCHES GIVE
THE LARGE BACK SCREEN
EXTRA STABILITY AGAINST
THE WIND.

Materials

12 lengths of rebar, each 6 feet long,
4 per screen (2 for post supports
and 2 to stabilize the rear)

1,000 pounds of fieldstone pavers

15 rolls of 20-gauge copper wire

Tools and Supplies

Fence-building tools and supplies listed on
page 103

Fine-tooth saw

Wire cutters

Sledgehammer

Cutting List

Description	Qty.	Material	Dimensions
Back screen posts	2	Bamboo poles	2½" x 90"
Left and right screen posts	4	Bamboo poles	2½" x 72"
Back screen stringers	2	Bamboo poles	2½" x 11'4"
Back screen stringers	2	Bamboo poles	1½" x 10'
Back screen stringer	1	Bamboo pole	1½" x 9'9"
Left and right screen stringers	4	Bamboo poles	2½" x 6'8"
Left and right screen stringers	4	Bamboo poles	1½" x 6'
Left and right screen stringers	2	Bamboo poles	1½" x 5½'
Back screen verticals	76	Bamboo poles, varying lengths	1½" x 76" to 88", one end cut just above the node
Left and right screen verticals	64	Bamboo poles	1½" x 72", one end cut just above the node
Sleeves for rebar supports	6	Bamboo poles	1½" x 4½'

Metric Equivalents

1"	2.5 cm	72"	1.6 m	100"	2.5 m	6'8"	2 m	
1½"	3.8 cm	76"	1.9m	2'	61 cm	9'9"	2.9 m	
2½"	6.4 cm	80"	2 m	4'	1.2 m	10'	3 m	
5"	12.7 cm	82"	2 m	4½'	1.35 m	11'4"	3.4 m	
12"	30.5 cm	88"	2.2 m	5½'	1.65 m	1,000 lbs.	454 kg	
52"	132.1 cm	90"	2.25 m	6'	1.8 m			

Instructions

1 Measure the site, and use the stakes and flagging to mark the placement of the stone ledges.

2 Within each of the three ledge sites, measure the placement of the rebar, which will support the posts. Center the rebar in the areas, 100 inches apart for the back screen and 52 inches apart for the left and right screens. Use the sledgehammer to drive two lengths of the rebar 2 feet into the ground.

3 Build the stone ledges around the rebar. Dry stack the pavers within the area, fitting and adjusting to set the stones securely and achieve an even outside edge.

4 Use the hammer and rebar to knock out the diaphragms of the lower 4 feet of each bamboo post.

5 On the front of the posts, measure and mark the points where the top and bottom

stringers (not the middle stringers) will be attached, 12 inches from the top and 5 inches up from the bottom. Cut the lines of the notches on the front of the poles. Saw to a depth of 1 inch, and use the hammer and chisel to knock out the in-between piece of bamboo. Slide the posts over the rebar.

6 Now attach the top and bottom stringers to the posts by lashing with copper wire.

7 Tightly lash the unnotched middle stringers to the backsides of the posts.

8 Now attach the vertical poles to the stringers, placing the closed-node ends up and using the level to check for straightness. Note that the front poles are wedged between the stringers. All

poles rest on the stone ledge.

9 Lash the poles by using one continuous roll of wire per stringer. Do this by wrapping each intersection with a basic cross tie (see page 44). Then, rather than cutting the wire when the lashing is complete, leave it intact to wrap the next pole. For the back screen, the grouping of poles is: five in front, six in back, five in front, six in back, five in front, eight in back, six in the center front, eight in back, five in front, six in back, five in front, six in back, and five in front. For the left and right screens, the grouping is: five in front, eight in back, six in the center front, eight in back, and five in front.

10 When placing the vertical poles in the back screen, use poles between 82 and 90

inches in length for the first group of poles on either side, then use poles 76 to 80 inches long for the central group. This will give a slight, inward dip to the staggered top edge.

11 Cut the ends of the stringers at an angle as shown.

12 Stabilize the screens in the back by angling two lengths of rebar between the screen and the ground. To hide the rebar and make the supports visible for safety, make bamboo "sleeves" by using a hammer and rebar to knock out the diaphragms of the 4½-foot poles. Slide the sleeves over the rebar. Hammer one end of the rebar into the ground, and wire the other end to the screen.

111

Yotsume Fence

S IMPLE IN CONSTRUCTION, THE *YOTSUME-GAKI* IS THE MOST COMMON BAMBOO FENCE IN JAPAN. ITS POLES GIVE SUPPORT TO LEANING SHRUBBERY, AND ITS OPEN SPACES GIVE PLANTS ROOM TO SPREAD. VARIATIONS ARE MANY. YOU CAN STAGGER THE TOPS FOR A COTTAGE FEEL, USE PAIRS RATHER THAN SINGLE POLES AS VERTICALS, OR USE TALLER POSTS AND STRINGERS FOR A HIGHER FENCE. THE FOLLOWING INSTRUCTIONS ARE FOR A 12-FOOT SECTION.

Materials

10 1⅝-inch drywall screws

50 1½-inch galvanized decking screws

Black hemp twine

Tools and Supplies

Wood stain, deep brown

Paintbrush

Fence-building tools and supplies listed on page 103

Rebar, 4 feet long or 2 feet long

Scissors

Cutting List

Description	Qty.	Material	Dimensions
Posts	2	Pressure-treated rounds	3" x 42"
Stringers	2	Bamboo poles	1½" x 12'4"
Verticals	25	Bamboo poles	1¼" to 1½" x 30", one end cut just above the node

Metric Equivalents

½"	1.3 cm	3"	7.6 cm	20"	50.8 cm	6'	1.8 m
1¼"	3.2 cm	5"	12.7 cm	30"	76.2 cm	12'	3.6 m
1½"	3.8 cm	6"	15.2 cm	42"	106.7 cm	12'4"	3.7 m
1⅝"	4.1 cm	12"	30.5 cm	4'	1.2 m		

Instructions

1 Brush the stain on the posts and let dry.

2 Measure the center points of the stringers. From the top of the posts, mark at 5 inches and 20 inches. Mark at 30 inches to indicate ground level.

3 Measure, mark, and cut notches in the posts, sawing to a depth of ½ inch and knocking out the in-between wood with the hammer and chisel.

4 Dig post holes and set the posts 12 inches into the ground. Trace the notches toward the front.

5 Place the stringers into the notches of the posts, extending their ends 6 inches beyond the posts. Drill pilot holes, and secure with the 1⅝-inch screws.

6 Select the strongest pole from the verticals to serve as the middle support. Use the rebar and hammer to knock out the diaphragms from the lower 12 inches of the pole.

7 Find the center of the fence, 6 feet in from each post. At this point, hammer in the rebar to a depth of 12 inches, leaving 12 inches above the ground. Place the middle support pole over the rebar.

8 Now you'll attach the verticals to the stringers. Measure and mark the top stringer every 6 inches, making 24 marks total. Next, add the first verticals next to a post. Use the level to check that the vertical is straight, drill pilot holes, and secure with the decking screws. Next, attach a vertical on the opposite side of the stringer. Continue adding verticals to alternate sides of the stringer.

9 Use the black hemp twine to tie the verticals to stringers, using traditional ties or simple lashings.

Opposite page: Yotsume fence and gate by the Bamboo Fencer.

Middle bamboo post

6'-0"

6"

5"

15"

42"

10"

24" Rebar

12"

Teppo Screen

with assistance from Allen Fowler and Julie Forkner

TEPPO SCREENS AND FENCES CONSIST OF GROUPS OF POLES PLACED ALTERNATELY ON BOTH SIDES OF THE STRINGERS. THIS MAKES THE FENCE LOOK DOUBLE-SIDED, ADDING DEPTH AND DIMENSION. IF YOU FACE A TEPPO SCREEN, THE VIEW BEYOND IS BLOCKED; LOOKED AT SIDEWAYS, THE VIEW OPENS. ALTHOUGH DESIGNED FOR A GARDEN, THIS SCREEN WORKS EQUALLY WELL ON A PORCH. PROVIDING TEXTURAL INTEREST AND PRIVACY, IT CAN MAKE ENTERING AND LEAVING YOUR HOUSE A PLEASURE.

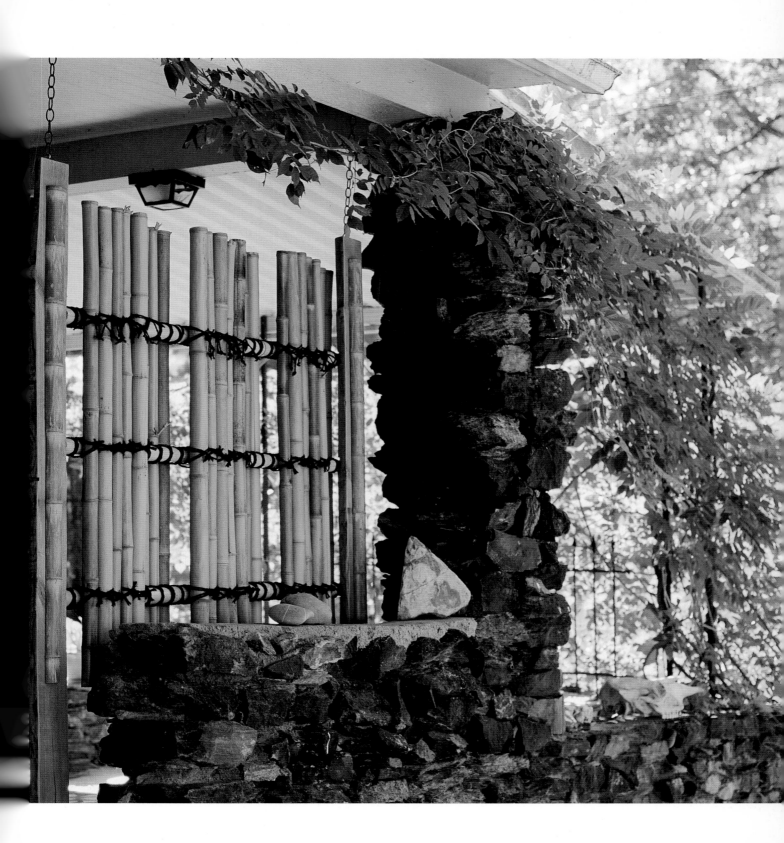

Materials

120 1½-inch galvanized decking screws

12 3-inch galvanized decking screws

Black hemp twine

10 1⅝-inch ringed nails

Tools and Supplies

Fence-building tools and supplies as described on page 103

Scrap piece of 2 x 4 lumber, 56 inches long

Scissors

Metric Equivalents

¾"	1.9 cm	18"	45.7 cm
1¼"	3.2 cm	30"	76.2 cm
1½"	3.8 cm	46"	116.8 cm
1⅝"	4.1 cm	52"	132.1 cm
3"	7.6 cm	56"	142.2 cm
4½"	11.4 cm	60"	152.4 cm
16"	40.6 cm	78"	198.1 cm

Cutting List

Description	Qty.	Material	Dimensions
Post facings	2	Split bamboo lengths	1¼" to 1½" x 52"
Posts	2	Pressure-treated half rounds	4½" x 78"
Stringers	3	Bamboo poles	1½" x 60"
Verticals	40	Bamboo poles	1¼" to 1½" x 52", one end cut just above the node

Instructions

1 Measure from the top of the post and mark the center points of the stringers at 16, 30, and 46 inches. Mark the base of the verticals at 52 inches and the ground line at 56 inches.

2 Measure, mark, and cut notches in the posts for the stringers, sawing to a depth of ¾ inch and knocking out the in-

Far left: A teppo-style screen gives privacy to the entrance of a home in Japan.

Left: Trimmed red willow shoots add texture to this bamboo screen.

<small>DESIGN AND PHOTO BY THE BAMBOO-SMITHS</small>

between wood with the hammer and chisel.

3 Dig holes in the ground for the posts, and set them 18 inches deep. Stabilize the posts with crushed rock or cement.

4 Secure the stringers to the posts. Make pilot holes, and use the 3-inch screws to attach.

5 Attach the 2 x 4 board at the 52-inch mark. It will serve as a temporary ledge on which you can rest the vertical poles while securing them to the stringers.

6 Attach the vertical poles by drilling pilot holes, then driving in the 1½-inch screws. Work back and forth, post to post, attaching groups of three and five poles on alternate sides of the stringers. Carefully plan and install the center set so no major gaps occur.

7 Use the black hemp twine to make decorative ties where verticals and stringers intersect.

8 Face the posts with the split bamboo lengths. Drill pilot holes and hammer in the ringed nails.

Figure 1

Bamboo Brush Fence

With assistance from
Sue Manderfield, Patty O'Keefe Hutton, and Scott Thompson

THE MASS OF BAMBOO BRUSH RISING FROM THE ORDERLY WALL OF POLES GIVES THIS SCREEN AN ENERGY WELL SUITED TO ITS PLACEMENT BETWEEN TREES. THE VARIETY SHOWN IS ARROW BAMBOO (*PSEUDOSASA JAPONICA*), BUT FEEL FREE TO EXPERIMENT WITH OTHER BAMBOOS! SELECT CULMS WITH BRANCHES AND BRANCHLETS THAT TIE EASILY INTO BUNDLES, AND SAVE TIME BY GATHERING CULMS THAT ARE DEAD AND HAVE DROPPED THEIR LEAVES.

Arrow bamboo sorted for use.

Instructions

1. Refer to figure 1 on page 123. Measure from the top of each post and mark the center point of the stringers at 4, 28, and 48 inches. Mark the ledge placement at 56 inches and the ground line at 60 inches.

2. Cut notches for the stringers by sawing 1 inch deep into the wood, and using the hammer and chisel to knock out the in-between wood.

3. Dig post holes and set the posts 18 inches into the ground.

4. Position the ledge between the posts, and use the level to make sure it's even. Drill the 1½-inch screws into the ends at an angle to secure the ledge into the posts.

5. Use the 1½-inch screws to attach two of the 1 x 2 lengths to the front and back of the ledge. These "lips" will keep the slender poles from slipping off the ledge.

6. Prepare the culms by shaking off the loose leaves. Use the garden pruners to cut off any leaves remaining on the branches; if left on, they will remain in the brush, darkening with mold and mildew.

7. Use the rope to make two parallel lines 7 feet apart on a large, flat surface. Lay out the

Materials

9 1½-inch galvanized decking screws

Binder twine

Black hemp twine

14 2-inch galvanized decking screws

12 2½-inch galvanized decking screws

21 1¼-inch ringed nails

Tools and Supplies

Fence-building tools and supplies listed on page 103

Garden pruners

Rope

Loppers

Scissors

Hedge clippers

Cutting List

Description	Quantity	Material	Dimensions
Posts	2	Pressure-treated 2½ x 3½ landscape timbers	6½'
Stringers	3	Pressure-treated 2 x 2 lumber	5½'
Ledge	1	Pressure-treated 1 x 4 lumber	5'
Front stringers and ledge lips	5	Pressure-treated 1 x 2 lumber	5'
Verticals	3 pick-up truckloads	Bamboo culms and branches	½" x 7'
Horizontals	2 split lengths	Bamboo poles	3" x 5' long

Metric Equivalents

½"	1.3 cm	3"	7.6 cm	56"	142.2 cm
1"	2.5 cm	3½"	8.9 cm	60"	152.4 cm
1¼"	3.2 cm	4"	10.2 cm	5'	1.5 m
1½"	3.8 cm	18"	45.7 cm	5½'	1.65 m
2"	5.1 cm	28"	71.1 cm	6½'	1.95 m
2½"	6.4 cm	48"	121.9 cm	7'	2 m

bamboo top to bottom, including branches, between the two lines. Use the loppers to cut off the culms that extend beyond the bottom line. These can be used in the bundles.

8 Assemble the trimmed poles into 15 piles, which when tied, will become the verticals. Since the brush makes the upper section wider, add the poles you cut in step 7 to make the lower sec-

tions wider. Use the binder twine to wrap them into tight bundles, each about 4 inches in diameter.

9 Attach the verticals to the stringers by positioning the base of a bundle on the ledge and wedging it between the lips. With more binder twine, tie the bundles temporarily to the stringers. Work back and forth from post to post, fitting one bundle close to the next, until they meet in the middle.

10 Now you'll attach the front stringers to hold in the brush. With a friend's help, use doubled binder twine to tie together the front stringer, the verticals, and the back stringer. Pull tight! The pole and brush verticals should be sandwiched securely between the stringers. Use the 2½-inch screws, driven in at an angle, to attach the front stringers to the posts.

11 Remove the temporary twine ties from the bundles, then even out the poles and branches.

12 Use the binder twine to temporarily tie the split horizontals to the front and back stringers. Mark the placement of ties.

13 Use the 2-inch screws to attach the horizontals to the front stringers where you've marked the ties.

14 Using doubled hemp twine, tie the traditional Japanese tie. Remove the temporary twine that held the horizontals in place.

15 Hammer in the ringed nails to attach the 5-foot poles over the front lip of the ledge. Use the hedge clippers to level the brush at the top.

4"

24"

60"

20"

7"

5"

18"

Figure 1

123

Bamboo Screen with Branch Posts

THIS SCREEN INTEGRATES KNOBBY, BRANCHED TREE LIMBS WITH THE STRAIGHT LINES AND UNPRONOUNCED NODES OF TONKIN BAMBOO. THE POSTS ARE PEELED TREE LIMBS, DRIED AND BRUSHED WITH A DEEP BROWN STAIN. YOU CAN TREAT THE VERTICAL POLES AS SHOWN OR CUT THEIR TOPS JUST ABOVE THE NODES.

Materials

Wood stain, dark brown

Polyurethane

12 2-inch galvanized decking screws

138 1¼-inch drywall screws

4 pieces of rebar, each 2 feet long

12 2½-inch galvanized decking screws

Black hemp twine

Galvanized wire

Tools and Supplies

Pruning saw

Scraping knife

Paintbrush

Power drill and assorted drill bits

Mason's cord

Fine-tooth saw

Scissors

Wire cutters

Metric Equivalents

1"	2.5 cm
1¼"	3.2 cm
1½"	3.8 cm
2"	5.1 cm
2½"	6.4 cm
3"	7.6 cm
12"	30.5 cm
14"	35.6 cm
18"	45.7 cm
28"	71.1cm
40"	101.6 cm
43"	109.2 cm
44"	111.8 cm
55"	139.7 cm
64"	188 cm
4'	1.2 m

Cutting List

Description	Qty.	Material	Dimensions
Posts	4	Sturdy tree limbs or branches	1½" x 64"
Stringers	3	Split bamboo lengths	2" x 55"
Verticals	46	Bamboo poles	1" x 43"

Instructions

1 Find fallen tree limbs or branches with dried bark that's easily removed from the wood. Scrape with the knife to obtain a clean surface. Brush on several coats of the brown stain, let dry, then seal with the polyurethane.

2 Measure from the top of each post, and mark the center point of each stringer at 14, 28, and 40 inches. Mark each post at 44 inches to indicate the bottom of the vertical poles. There will be 12 inches between the verticals and the ground.

3 On a flat surface, lay out the tree posts parallel to each other, 18 inches apart. Position the stringers across the posts. Drill pilot holes, then drive in the 2-inch screws. You've now formed the back of the fence framework.

4 Turn the framework over to the front side. Drive in 1¼-inch screws at the 44-inch marks on the posts, and stretch the

mason's cord between the screws. This will be the guide to keep the base of the poles even. Place the verticals between the posts, adjusting for a good fit, especially near the branches. Some may need to be cut to accommodate protruding branches. Use the drill to make pilot holes, then use the 1¼-inch screws to attach the verticals to the stringers.

5 Lay the three split horizontals across the screen over the line of screw heads. If necessary, trim the ends to fit up against the end poles. Mark the position of the ties. At those points, drill pilot holes and use the 2½-inch screws to attach the horizontals to the verticals and stringers.

6 To install the screen, mark the position of the posts at the site. At these points, hammer the rebar 12 inches into the ground, then place the posts up against the rebar. Use the wire to wrap the posts and rebar together to stabilize the screen.

7 Use the black hemp twine to make ornamental ties if desired.

14"

28"

40"

18"   18"   18"

44"

12"

24" Rebar

Reed Fence with Bamboo-Faced Stringers

Designer, Carolyn Brooks, with assistance from Carol Stangler

THIS IS THE PERFECT FENCE IF YOU WANT A FEELING OF PRIVACY WITHOUT CREATING A VISUAL BARRICADE BETWEEN YOU AND YOUR NEIGHBORS. THE PANELS, MADE OF REED FENCING STRETCHED ACROSS POSTS AND STRINGERS, MOVE IN AN IRREGULAR ZIGZAG TO AVOID BUSHES AND STUMPS. THE 42-FOOT FENCE SHOWN IN THE PHOTOGRAPH HAS A SOFT, ORGANIC FEEL, AS IF IT HAS ALWAYS BEEN PART OF THE LANDSCAPE. THE TEXTURE OF THE REED COMPLEMENTS BAMBOO PERFECTLY.

Instructions

1 Measure from the top and mark the center point of the stringers at 24 inches and 48 inches. Mark the ground line at 57 inches.

2 Dig post holes, and set the posts 15 inches into the ground. Fill with crushed rock or cement.

3 Use the 2½-inch screws to attach the stringers to the top of the posts. Next, fit the middle and bottom stringers between the posts, and attach with the 1⅝-inch screws.

4 Unroll the reed fencing on the ground and flatten it. Don't worry about any irregularities; they lend charm to the material. With

help from a friend, move the fencing up against one side of the fence.

5 Now you'll attach the reed fencing to the posts and stringers. Secure the fencing to an end post with the 10-inch pieces of wire, and straighten and align to your satisfaction. Then use the 8-inch lengths to attach

Materials

78 2½-inch galvanized decking screws

72 1⅝-inch drywall screws

72 1¼-inch drywall screws

20-gauge galvanized wire, cut into 8- and 10-inch lengths

Tools and Supplies

Fence-building tools and supplies listed on page 103

Bull-nose pliers

Wire cutters

A helpful friend

Cutting List

Description	Qty.	Material	Dimensions
Facings on large panel stringers	6	Split bamboo lengths	1½" x 7'8"
Facings on smaller panel stringers	3	Split bamboo lengths	1½ x 4'10"
Stringers for 5 large panels	15	Pressure-treated, 2 x 2 lumber	7'8"
Stringers for smaller panel	3	Pressure-treated, 2 x 2 lumber	4'10"
Posts	7	Pressure-treated timber rounds	3" x 6'
Verticals	6	Rolls of reed fencing	4 x 8'

Metric Equivalents

1¼"	3.2 cm	8"	20.3 cm	57"	144.8 cm	8'	2.4 m		
1½"	3.8 cm	10"	25.4 cm	4'	1.2 m	42'	12.6 m		
1⅝"	4.1 cm	15"	38.1 cm	4'10"	1.4 m				
2½"	6.4 cm	24"	61 cm	6'	1.2 m				
3"	7.6 cm	48"	121.9 cm	7' 8"	2.3 m				

the fencing to the stringers, and the 10-inch lengths to attach the fencing to the posts. Work as a team with your friend. The person on one side will adjust the fencing, bend the wire, and pass it through to the other person. That person will use the pliers to pull and twist the ends together. Use the wire cutters to cut the ends close to the twist, and tuck between the reed. When you reach the end of one roll, wire another roll to it or create a slight overlap. When you reach the end post, wrap the reed around it and secure with wire.

6 Drill pilot holes, and use the 1¼-inch drywall screws to attach the split bamboo to the stringers. Take care not to drive the screw in too far, or the bamboo may crack.

Chapter 9

Gates, Railings, and Outdoor Structures

ATES SERVE AS OPENINGS AND PASSAGEWAYS THAT SIGNAL A CHANGE FROM ONE SPACE TO ANOTHER. AS BOUNDARY MARKERS, ENTRANCES, AND EXITS, THEY EVOKE A SENSE OF MYSTERY AND ADVENTURE. WHEN BAMBOO IS USED TO CONSTRUCT OR EMBELLISH A GATE, ITS ROUNDED, CYLINDRICAL WALLS IMPART A COMFORTING PRESENCE THAT ENHANCES THE PLEASURE OF OPENING, PASSING THROUGH, AND CLOSING A GATE. WHEN DESIGNING A GATE THAT WILL INCORPORATE BAMBOO, CONSIDER ITS SURROUNDING ENVIRONMENT AND DESIGN THE GATE TO CONNECT AND HARMONIZE WITH IT.

Gate and fence connect seamlessly by incorporating similar materials and design motifs.
DESIGN AND PHOTO BY ROBERT SMALL

Left: The translucent panels of this wall dividing system are framed in bamboo.

DESIGN AND PHOTO BY MICHAEL ELLISTON

Right: Split bamboo is arranged and bent to form an entrance arch at the Prafrance Gardens, France.

PHOTO BY ADAM AND SUE TURTLE, ©*TEMPERATE BAMBOO QUARTERLY*, USED WITH SPECIAL PERMISSION.

Left: Bamboo railing at the Brooklyn Botanic Garden.

Lower left: Cedar and bamboo fence and gate.

Right: An open bamboo door invites visitors to a meditation garden in Nashville, Tennessee.

Far right: Bamboo railings are used along steps to guide worshippers to a temple in Japan.

Below: Lengths of cedar frame this gate of reed fencing and split bamboo.

LONG, STRAIGHT CULMS of bamboo are a natural material for handrails and railings. Since the purpose of a handrail is to provide support, it's important to firmly set posts in the ground, to use strong poles, and to securely fasten the poles to the posts. Use poles easily grasped by the hand; a 2-inch (5.1 cm) diameter works well. The Bamboo Handrail project on page 146 is a good example of these principles.

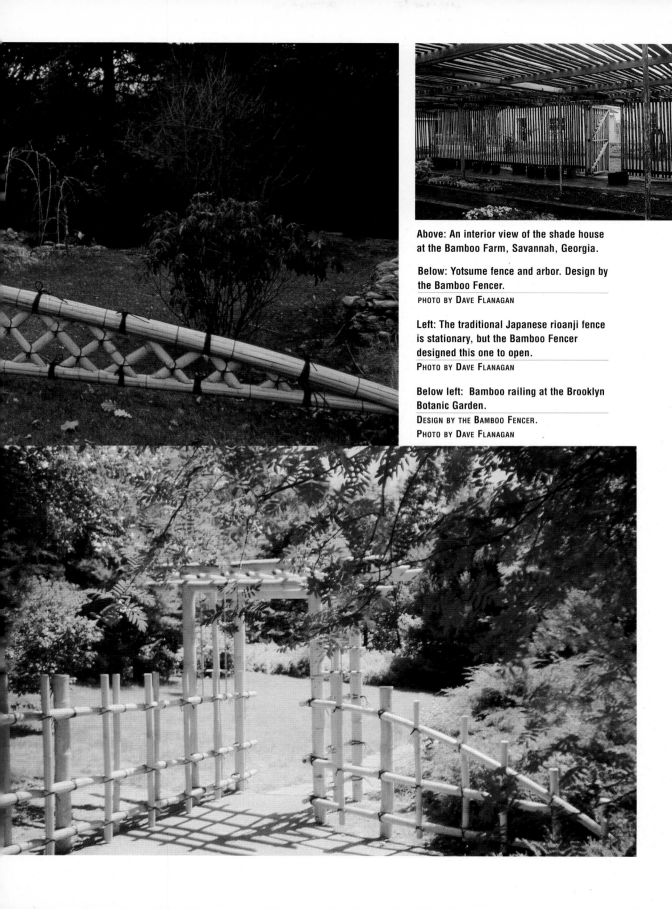

Above: An interior view of the shade house at the Bamboo Farm, Savannah, Georgia.

Below: Yotsume fence and arbor. Design by the Bamboo Fencer.

PHOTO BY DAVE FLANAGAN

Left: The traditional Japanese rioanji fence is stationary, but the Bamboo Fencer designed this one to open.

PHOTO BY DAVE FLANAGAN

Below left: Bamboo railing at the Brooklyn Botanic Garden.

DESIGN BY THE BAMBOO FENCER.

PHOTO BY DAVE FLANAGAN

Sunrise and Mountain Bamboo Mosaic Gate

With assistance from Allen Fowler, Patty O'Keefe Hutton, and Lewis Lance
Steel bird head ornaments forged by Joe Babb

THIS GATE ADAPTS THE RUSTIC FURNITURE TECHNIQUE CALLED *MOSAIC WORK*, IN WHICH TWIGS ARE NAILED IN PATTERNS ON WOOD SURFACES.

THE GATE IS FACED WITH A VARIETY OF BAMBOO LAID OUT IN A HER-RINGBONE PATTERN. ITS TEXTURES, COLORS, AND ANGLES CREATE A TABLEAU OF MOUNTAINS, VEGETA-TION, AND SUN RAYS, A JUXTAPOSI-TION THAT CREATES SPONTANEITY AND MOVEMENT.

ding marks, and with chalk or pencil, draw an outline around each one, but don't attach them to the post yet.

4 Place the gate on your work surface. Mark the center

screw holes of the hinges on the side that will swing from the gatepost. Measure from the top of the gate, and mark as shown in figure 1. Use the drill to attach the hinges to the gate with their supporting screws.

Instructions

1 Stain the gateposts with the brownish black stain, applying two coats with the brush. Let dry between coats.

2 Dig the holes for the posts, and set 18 inches into the ground. Stabilize with cement.

3 Referring to figure 1, mark the placement of the hinges on the post from which you'll hang the gate. (The gate shown in the photo on page 137 is hung from the left post; hang yours from the right post if desired.) Measuring from the top, mark the center screw holes for the hinges at 2¼, 19¼, and 41¼ inches. Position the hinges over the correspon-

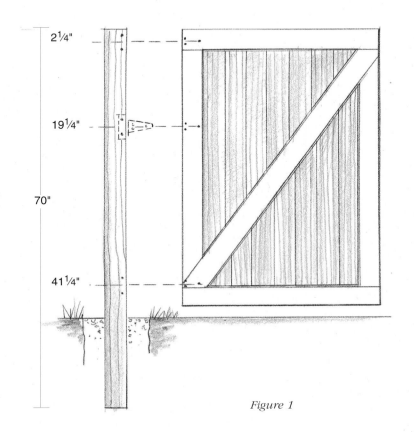

2¼"

19¼"

70"

41¼"

Figure 1

Materials

Wood stains, brown-black and rosewood colors

Polyurethane

3 4-inch galvanized T-hinges with supporting screws

Gate, 48 inches high and 37 inches wide, 1½ inches thick, constructed of 1 x 4 boards

Wood dye in yellow-gold

Gate handle and attaching screws

150 assorted drywall and galvanized decking screws, 1 to 2½ inches long

25 1⅜-inch brass wood screws

10 1¼-inch ringed nails

Post ornaments (optional)

Tools and Supplies

Fence-building tools and supplies listed on page 103

Chalk or pencil

Paintbrush

Scraping knife

Sanding sponge

Fine-tooth saw

Metric Equivalents

⅜"	9.5 mm
½"	1.3 cm
¾"	1.9 cm
1"	2.5 cm
1¼"	3.2 cm
1⅜"	3.4 cm
1½"	3.8 cm
2"	5.1 cm
2¼"	5.7 cm
2½"	6.4 cm
3"	7.6 cm
4"	10.2 cm
5½"	14 cm
10"	25.4 cm
12"	30.5 cm
15"	38.1 cm
18"	45.7 cm
19¼"	48.9 cm
19½"	49.5 cm
22½"	57.2 cm
37"	94 cm
37½"	95.3 cm
39"	99.1 cm
40"	101.6 cm
41¼"	104.7 cm
44¼"	112.4 cm
53"	134.6 cm

Cutting List

Description	Qty.	Material	Dimensions
Gateposts	2	Pressure-treated 4 x 4 posts	70"
Gate ornament pedestals (optional)	2	Bamboo pieces	2" long, cut just above the nodes
Mountain mosaic material	35 to 50	Black, moso, golden, and/or henon bamboo in whole and split pieces*	10" to 40", ¾" to 3" diameter
Sun rays	11	Zigzag-shape pieces of golden bamboo (phyllostachys aurea)	12" to 19½"
Orb of sun	1	Curved bamboo piece	15" separating 2 ends, 1" diameter
Mosaic parts	5 to 10	Tamarack or other long, thin branches	⅜" x 10" to 39"

5 Draw the top line of the mountains on the gate, using the T-square to create the 45° angles for the peaks. This line separates the mountain area from the sky area. Brush on two coats of the brownish black stain in the mountain area and two coats of rosewood stain on the sky area. Let dry between coats, and after it is thoroughly dry, apply two coats of polyurethane over the entire surface.

6 Begin the mountain mosaic by laying out the defining bamboo pieces, as shown in figure 2. Below these, choose whole and split pieces and mix varieties, colors, and diameters to fill in the design and suggest differences in the landscape. Working from the mountaintops down, saw the pieces to length, sanding the ends and fitting them at right angles to each other. Mark and cut the outer ends on the diagonal so the bamboo stops ½ inch within the edges of the gate. Continue setting in lengths of bamboo and the tamarack branches until the mountain scene is complete.

7 Attach the bamboo pieces to the gate. Make pilot holes and drive in the screws, using various screw lengths to attach pieces as needed. Use the hammer and ringed nails to attach the narrow bamboo poles and the tamarack branches.

8 Scrape the zigzag lengths and the piece of curved bamboo to remove the waxy outer coating. Brush on two coats of the yellow-gold wood dye, letting dry between coats. Brush on a coat of polyurethane and let dry.

9 Position the curved piece of bamboo between the mountain peaks, and arrange the zigzag rays above it. Use the drill to make pilot holes, and attach the sun and sun rays with the brass wood screws. Position the gate handle and attach with screws.

10 Bush the bamboo mosaic with two coats of polyurethane. Let dry between coats.

11 Hang the gate by matching the hinges against the outlines you marked in step 3, checking to make sure the gate swings in the desired direction and that there is enough space for it to open and close. Screw the hinges to the post. If your handle has a catch, screw it in place on the opposite post.

12 Attach the ornaments to the 2-inch pieces of bamboo by drilling pilot holes through the walls and driving in the 1-inch screws. Use the drill to make two angled pilot holes through the base of the bamboo pedestals into the post, then hammer in ringed nails.

Mosaic lengths should end ½" from edge of gate.

Figure 2

Reed Fencing over Wire Gate

Designer, Carolyn Brooks,
with assistance from
Carol Stangler

I N THIS PROJECT, A 1950S-ERA METAL AND WIRE GATE IS INGENIOUSLY TRANS-FORMED BY ATTACHING REED FENCING TO ITS FRONT AND BACK. A DRIFT-WOOD HANDLE, CROOKED LENGTHS OF BAMBOO, AND SHELL ORNAMENTATION EVOKE IMAGES OF THE BEACH. THE GATE SHOWN HERE MEASURES 38 INCH-ES WIDE AND 50 INCHES HIGH, BUT YOU CAN ADAPT THE DIMENSIONS TO FIT A GATE OF DIFFERENT PRO-PORTIONS.

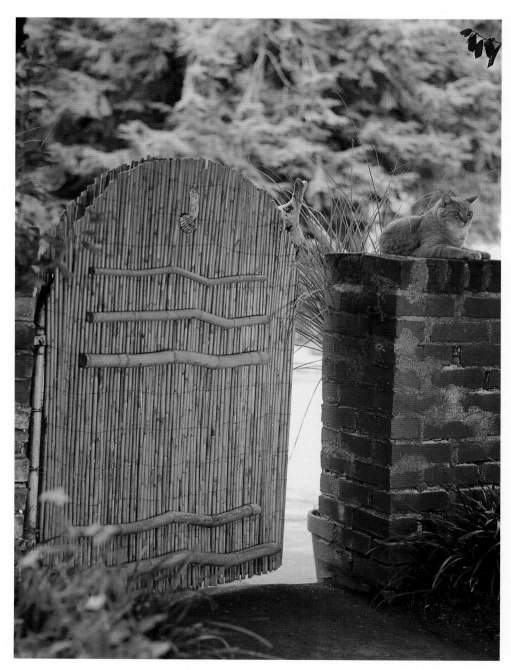

Instructions

1 Remove the wire gate from its hinges, and lay it on top of one piece of reed fencing. Using the marker, trace the gate outline (top curve and side edges) onto the fencing. Repeat with the other piece of fencing.

2 Clamp the reed fencing to a work surface, and use the jigsaw to saw along the curved lines, cutting through the reed and the wire that binds the reeds together.

3 Set the wire gate on top of and between the sawhorses. Position a piece of the reed fencing over the gate. Fold the sides of the reed under so it covers the metal frame but doesn't extend over the edge.

4 Put on the work gloves, and use the wire cutters to cut 5-inch lengths of the galvanized wire. Bend the wire into U shapes. Attach the fencing first to the metal frame, then to the wire strands within the frame. Do this by inserting the bent wire through the fencing and around the metal frame or the strands of wire within the frame. Twist tightly with the pliers to secure. Trim excess wire, and tuck the ends inward.

5 Repeat steps 3 and 4 to attach the other piece of fencing on the gate's reverse side.

6 Working on one side, lay the zigzag bamboo pieces on top of the fencing in a horizontal pattern of three lengths at the top and two below. Cut 10-inch lengths of wire and bend them in a U shape. Use them to secure the lengths of bamboo by passing them through both panels of reed fencing and coming out on the other side. With the pliers, twist tightly, cut with the wire cutters, and tuck the ends. Repeat on the reverse side.

7 Reattach the gate to its hinges. Position the piece of driftwood between the reed fencing so it rests securely on the gate's metal frame slightly off to the side. Use the silicone glue to attach it to the frame, applying a generous amount to the handle and metal, then clamp in place for 24 hours. (Silicone is a good choice because its thick gel adheres to rough surfaces, and it expands and contracts with warm and cold weather.) Secure the handle further by lashing with the wire.

8 Use the glue to attach the shells to the fencing.

Materials

Wire gate

2 pieces of reed fencing, each 40 inches wide by 50 inches high

Handle-shaped piece of driftwood

Shells

Metric Equivalents

1"	2.5 cm
5"	12.7 cm
10"	30.5 cm
34"	86.4 cm
38"	96.5 cm
40"	101.6 cm
50"	127 cm

Tools and Supplies

Fine-tip permanent marker

Quick-release clamps

Jigsaw with metal blade

Pair of sawhorses

Work gloves

Bull-nose pliers

Wire cutters

Fine-tooth saw

Silicone glue, clear

20-gauge galvanized wire

Cutting List

Description	Qty.	Material	Dimensions
Ornamentation	10	Zigzag lengths of golden (*Phyllostachys aurea*) bamboo	1" x 34"

Clockwise from top: Bamboo poles are attached on the diagonal to create this fence at the Bamboo Farm near Savannah, Georgia.

This fence of split bamboo is capped, with horizontals bordering the top and bottom.
DESIGN BY FRANK LINTON

Bamboo fences and screens create privacy outside a residence in Japan.

Nanako Border

THIS LOW, CURVED FENCE IS FREQUENTLY SEEN BORDERING PATH-
WAYS IN JAPANESE PARKS AND TEMPLE GROUNDS. AS SHOWN
HERE, IT ALSO MAKES A DELIGHTFUL BORDER FOR FLOWER BEDS.
SIMPLE TO CONSTRUCT AND INSTALL, A NANAKO IS TRADITIONALLY TIED WITH
BLACK HEMP TWINE AT THE INTERSECTIONS OF ITS HOOPS. THESE DIRECTIONS
PROVIDE ENOUGH HOOPS FOR A 34-FOOT WALKWAY.

Instructions

1 Use the splitting knife and mallet to split the poles into 10 halves. Hammer out the diaphragms, then split each length in half again, making a total of 20 splints, each approximately ⅝ inch wide.

2 Use the splitting knife to remove the pulp of the inner wall from 17 of the splints.

3 Use the knife or shaver to smooth the inner side of the splints. Test for the proper thinness and evenness by holding one end in each hand and bringing the hands together to form a hoop about 28 inches across. If the bamboo makes a gentle curve, good; if the splint bends at an angle, take off more pulp along the inner wall.

4 Along your walkway or flower bed, measure and mark 28-inch intervals with the stakes. The stakes indicate where the center of two splints will overlap.

5 Install the border by bending each splint into a hoop with the ends 28 inches apart. Position each end 2 inches on either side of the center mark. Insert the ends 4 inches into the ground. Continue installing the splints and overlapping their ends.

6 When all the splints are installed, adjust their heights so the tops are even.

Materials, Tools, and Supplies

5 bamboo poles, each 50 inches long, 1½ inches in diameter*

Splitting knife

Mallet

Hammer

Knife or shaver

Measuring tape

Stakes

Sanding sponge

*These will be split to make 20 hoops.

Metric Equivalents

⅝"	1.6 cm
1½"	3.8 cm
2"	5.1 cm
4"	10.2 cm
28"	71.1 cm
50"	127 cm
34'	10.4 m

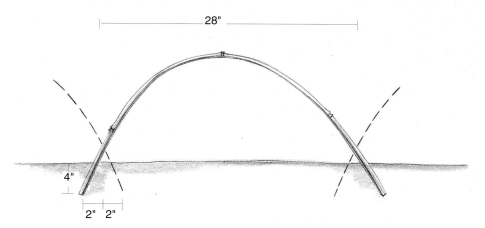

Bamboo Handrail

with assistance from Allen Fowler

ANDRAILS ARE INVALUABLE FOR HELPING PEOPLE NEGOTIATE STEPS AND WALK-
WAYS. THIS SIMPLE DESIGN USES A CONTINUOUS LINE OF POLES ATTACHED TO THE
TOPS OF POSTS AND CAN BE ADAPTED TO FIT ANY WALKWAY. DIRECTIONS FOR
CREATING THE ROW OF CURVED BAMBOO SPLINTS ARE FOUND IN THE NANAKO BORDER PROJECT
ON PAGE 144.

Materials

Wood stain, brownish black

2 4½-inch lag screws with washers and nuts

2 handrail brackets

4 1-inch drywall screws

8 3-inch galvanized decking screws

4 1½-inch galvanized decking screws

Water sealer

Metric Equivalents

⅛"	3 mm	18"	45.7 cm
¼"	6 mm	30"	76.2 cm
½"	1.3 cm	36"	91.4 cm
1"	2.5 cm	42"	1 m
1½"	3.8 cm	54"	1.35 m
2"	5.1 cm	79"	2 m
3"	7.6 cm	96"	2.4 m
4"	10.2 cm	6'	1.8 m
4½"	11.4 cm		

Tools and Supplies

Fence-building tools and supplies listed on page 103

Handsaw or power saw

Bricks

Paintbrush

Wrench

Countersink bit

Fine-tooth saw

Rebar, 6 feet long

Wood glue

Sanding sponge

Cutting List

Description	Qty.	Material	Dimensions
Stair post reinforcements	2	Pressure-treated round posts	3" x 42"
Handrail along steps	1	Bamboo pole*	2" x 96"
Handrail supports	7	Pressure-treated round posts	3" x 54"
Walkway railings	4	Bamboo poles**	2" x 79"

*This pole should have one end cut just above a node, and the other end as close to a node as possible.

**Select smooth-textured poles. Two should have one closed node end to serve as the beginning and end of the railing. The other two poles can have open ends; they'll be positioned in the middle.

Instructions

Making the Posts

1 Referring to figure 1, mark and cut the posts to support the railings along the walkway. On the tops of five of the 54-inch posts, make parallel marks ½ inch in from each edge. On the same two sides of the posts, measure down 1½ inches and mark. Starting at the top, saw at an angle, cutting off the triangular pieces. The bamboo poles will rest on the 2-inch-wide flat tops of the posts.

2 Now you'll cut the tops of the posts to be used for the handrail at the stairs. First, determine the angle at which you'll make the cut; it should be equivalent to the angle of the steps. This is most easily done by digging the holes for the posts with a post-hole digger and inserting the posts 18 inches into the ground. Secure them temporarily with the

bricks using a level to check for straightness. Measure up 36 inches from the steps, and mark the posts. Clamp, or have a friend hold, the pole at the mark on one post while you hold the pole at the mark on the other post. This should give you an angle parallel to the slope of the stairs. Mark this angle onto the posts. Remove the posts, and cut the tops at that angle with the hand- or power saw.

3 Brush the posts with two coats of stain. Let dry between coats.

4 Measure and mark the placement of the posts along the walkway. Use the post-hole digger to dig holes along the walkway, and insert posts to a depth of 18 inches. Set the posts in gravel and tamp firmly. Be sure to use a level to achieve straight posts with level tops.

5 Referring to figure 2, dig holes for the two 42-inch posts along the stairs in similar fashion, but create additional support by digging a wider hole and setting the 30-inch posts beside them. Set the poles in the holes so the downward cut is facing you as you stand at the base of the stairs. Join the two posts with the lag screws, first drilling out the hole. Use the wrench to turn the screw, and secure the ends of the screws

with the washers and nuts. Set the posts in concrete.

Making the Handrail along the Steps

1 Attach the handrail brackets to the two posts beside the steps. On the sides of the posts that face the stairs, measure 4 inches down from the top and make a line across, parallel to the step below. Next, determine the center of the post and draw a vertical line. At the point where the lines intersect, position the top of the handrail bracket plate, and mark the outline on the posts.

2 Since the post is round, you'll need to make notches by removing wood from each post where the plates will be attached. Use the handsaw to cut the top and bottom lines of the plate outline to a depth of ¼ inch, as shown in figure 3. With the hammer and chisel, knock out the wood in between the lines. Repeat with the other post. Position an arm bracket in each notch, making sure the fit is stable. Use the drill and 1-inch screws to attach the brackets to the post.

3 Position the handrail pole across the arm brackets so the closed-node end is at the base of the stairs. Rotate it to find the firmest fit. Mark the placement

of the attachment plates by outlining the plates on the underside of the pole.

4 Use the fine-tooth saw to cut out the curve of the bamboo at the attachment plates, so the handrail brackets will be held firmly. Do this by carefully sawing along the top and bottom lines of the outline to a depth of about ⅛ inch. Use the hammer and chisel to carefully knock out the bamboo in between the lines.

5 Apply the wood glue to the surfaces of the pole and the bracket. Join them by drilling pilot holes and sinking in 1-inch screws. Clamp into place and let dry overnight. Any slight wiggle of

Top view attaching handrail poles to post

Midpoint

½"

½"

Cutting the posts

Figure 1

the handrail can be eliminated by inserting thin bamboo pegs between the brackets and the handrail.

Making the Railing along the Walkway

1 To help prevent splitting, use the hammer and rebar to knock out all the diaphragms from the three middle poles. Knock out the diaphragms in the other two poles, except for the end nodes.

2 Now you'll attach the poles to the tops of the posts along the walkway. First use the T-square to mark a line on the flat top of each post, perpendicular to the walkway, to indicate the mid-point. Refer again to figure 1.

3 Starting with the post at the top of the stairs, take one of the poles with the closed-node end and position it so the pole overhangs the stairs by 2 inches. Position the other end on the second post, rotating the pole to find the firmest fit between the two posts.

4 Notice that the pole end rest-ing on the second post extends at an angle slightly beyond the line of the midpoint. Since the railing needs to make a curve at this point, the end must be cut along the midpoint line. Use the permanent marker to transfer the midpoint line to the bamboo, and use the fine-tooth saw to cut along the line. Use the sanding sponge to smooth the edges.

Figure 2

Figure 3

Figure 4

5 As shown in figure 4, secure the poles to the post with the 3-inch screws, first drilling pilot holes and using the countersink bit to make crevices for the screw heads. Be very careful not to sink the screws too deeply into the bamboo, or the pole may crack.

6 Add the next pole, which will be open at each end. Rotate to find the firmest position. Transfer the midpoint line from the second post onto that end of the bamboo pole, then cut, sand, and rest it against the first. The joint of the two poles should be continuous so that a user's hand glides smoothly across. If the height differs, rasp the bottom side so it rests flat.

7 Install the remaining poles. The railing should end with a closed node. Brush the posts and rail with the water sealer, and let dry.

Al Fresco Shower Stall

with assistance from Charlie Pope

THE WALLS OF THIS DELIGHTFUL OUTDOOR SHOWER ARE MADE FROM GROUPS OF BAM-
BOO POLES ALTERNATING WITH BRUSH FOR TEXTURAL CONTRAST. THE UPRIGHT POLES
ECHO THE PARALLEL LINES OF THE DECK FLOORING, AND THE STRUCTURE CAN SUP-
PORT CLIMBING VINES. AS A BONUS, THE BACK WALL IS LEFT OPEN TO GIVE A VIEW OF THE WOODS
FROM INSIDE THE STALL. YOU CAN HAVE PLUMBING INSTALLED TO FIT THE SPECIFICATIONS OF THIS
PROJECT, OR ALTER THE DIMENSIONS TO FIT EXISTING PLUMBING.

Instructions

1 Refer to figure 1. Measure and mark the placement of the 2 x 2 stringers on the posts of the side and back walls. Measuring from the top of the posts, mark the centerpoints of the stringers at 25½, 41½, and 57 inches. Mark at 75 inches to indicate the bottom line of the bamboo and brush verticals.

2 Center the stringers over the marks on the posts with 1½ inches of each stringer extending over the post on each side. Use the level and T-square to ensure the stringers are even. Use the 2½-inch screws to secure the stringers to the posts.

3 Use the binder twine to tie the brush into bundles approximately 2 to 3 inches in diameter.

4 Now you'll attach the groups of vertical poles to the stringers on the two side walls. In each wall, use two groups of four and one group of three verticals. Drill pilot holes, and use the 1⅝-inch screws to secure the verticals. Use the binder twine to tie the bundles of brush onto the stringers between the groups of poles.

5 Position a split bamboo horizontal over each line of screw heads. Using the binder twine, tie the horizontal in place at both ends so it fits tightly against the cedar posts. Use the 2½-inch screws to attach the horizontals to the stringers. You now have stringers, brush and bamboo poles, and a split horizontal installed on top of each other in that order, all packed tightly together between the 4 x 4 cedar foot posts.

Side wall Back wall Front

Figure 1

Materials

4-sided shower stall frame, 8 feet tall, 3 feet wide, constructed of 4 x 4 cedar posts, with plumbing installed

105 1⅝-inch drywall screws

48 2½-inch galvanized decking screws

2 3-inch galvanized strap hinges with supporting screws

4 1¼-inch ringed nails

1¼-inch decorative brass wood screw

Hook and eye latch

Tools and Supplies

Measuring tape

Fine-tip permanent marker

Level

T-square

Power drill and assorted drill bits

Binder twine

Fine-tooth saw

Splitting knife

Mallet

Sanding sponge

Hedge clippers

Metric Equivalents

½"	1.3 cm	2½"	6.4 cm	32"	81.3 cm
1"	2.5 cm	3"	7.6 cm	36"	91.4 cm
1¼"	3.2 cm	4"	10.2 cm	39"	99.1 cm
1½"	3.8 cm	5"	12.7 cm	41½"	75.4 cm
1⅝"	4.1 cm	7"	17.8 cm	57"	144.8 cm
1¾"	4.4 cm	8"	20.3 cm	58"	147.3 cm
2"	5.1 cm	25½"	64.8 cm	74"	188 cm

Cutting List

Description	Qty.	Material	Dimensions
Stringers	11	Pressure-treated 2 x 2 lumber	39"
Door and side wall verticals	35	Bamboo poles, tops cut above a node	1" to 1¼" x 58"
Back wall staggered verticals	16	Bamboo poles, tops cut above a node	1" to 1¼" x 32" to 58" long
Vertical brush	1 pick-up truckload	Arrow bamboo (*Pseudosasa japonica*) culms and branches	58"
Horizontals	11	Split bamboo lengths	1½" to 1¾" x 36"
Door latch fixed part	1	Bamboo length	2" x 5"
Door latch moveable part	1	Flat bamboo splint	2" x 7"
Latch side gate frame	1	Pressure-treated 2 x 2 lumber	32"
Hinge side gate frame	1	Pressure-treated 2 x 4 lumber	32"

TOP VIEW CEDAR FRAME, DOOR FRAME, AND UPPER STRINGERS

2 x 4 Upper frame

4 x 4 Cedar post

2 x 2 Stringers

36"

2 x 4

2 x 2

Figure 2

FIXED PART OF DOOR LATCH

½" 1"

3"

2"

1½" ½"

Figure 3

6 To make the back wall, repeat steps 2 through 5, but cut and position the lengths of vertical poles to create a semicircular opening.

7 Assemble the door on a flat surface. As shown in figure 2, position the stringers across the 2 x 4 and 2 x 2 which make up the sides of the door frame. The side pieces should be 36 inches apart, with the stringers extending 1½ inches over the frame. Make sure all pieces are level and square. Use the 2½-inch screws to attach the stringers to the frame.

8 Repeat steps 2 through 5 to attach the verticals and horizontals to the door.

9 Use the supporting screws to attach the hinges to the outside edge of the 2 x 4 door frame, centering one hinge 8 inches from the top and the other 8 inches from the bottom.

10 Now you'll hang the door. Mark the center point of the hinges on the supporting post, measuring 4 inches down from the top of the post and 4 inches up from the bottom. Check the position of the hinges to make sure the door opens in the proper direction and will open and close easily. Screw the hinges to the post with the supporting screws.

11 To make the stationary part of the bamboo latch, refer

to figure 3. Use the splitting knife and mallet to remove ½ inch from a side of the 5-inch piece of bamboo. From the split side, saw across ½ inch, 2 inches from the base. Use the splitting knife to remove the splints of bamboo above the cuts, creating a lip for the latch to rest upon. Use the sanding sponge to smooth the edges.

12 Position the fixed part of the latch on the 2 x 2 side frame piece. Line up the tip of the latch with the middle horizontal. Use the drill to make pilot holes in each corner of the latch base, and hammer in the panel nails.

13 Make the moveable part of the latch by sanding the edges of the 2 x 7-inch flat splint. Drill a hole slightly larger than the diameter of the brass screw through the splint, centered 1 inch from the right edge. Position the moveable piece across from the fixed part of the latch; when closed, its end should line up with the middle horizontal. Check to make sure the latch opens and closes easily. Use the drill to drive in the brass screw.

14 Remove the twine from the horizontals and the brush bundles. Even out the slender poles and branches. Use the hedge clippers to level the top.

15 Install the hook and eye inside the stall for privacy.

Acknowledgments

The Craft and Art of Bamboo crystallized around my vision of bamboo as a beautiful, versatile material now readily available to craftspeople and gardeners. Toward this end, I immersed myself in all things bamboo, with the desire to contribute toward a new field of contemporary bamboo art. Pursuing this personal vision, I walked the path with great faith, great doubt, and great determination. Fortunately, with the knowledge and enthusiasm from the many artists, growers, and friends of bamboo I encountered, my vision expanded and solidified. To this steady stream of acquaintances who contributed to the book, as well as to my family and friends who supported my efforts, thank you each and every one.

Among those who stand out for catalyzing ideas and energy in the process are Gail and Brian McCarthy, Kari Lønning, Lori Feig-Sandoval, and Francisco Plaza. To the trio who gave me moral support during the trials of putting the vision into words, thank you Toni Giardino, Allen Fowler, and Mary Thompson. As always, deep gratitude to The Bird Tribe, the joyful and ever-present elementals of the unseen world who guide and enhance my creative life.

Bows of respect to my bamboo-crafting teachers, Brother Konomo Utsumi, Takei Akinori, Reed Hamilton, and Doug Lingen, as well as to the leadership of the American Bamboo Society for their work around the globe in collecting and disseminating knowledge of this most amazing plant. Among

those leaders, a personal thank-you to Susanne Lucas and Karen Vohman for their generosity and energy, and to all those who came before me, growing and crafting bamboo for decades. Of these pioneers, special thanks to Adam and Sue Turtle, Frank Linton, Dave Flanagan and Steve Ray, each of whom in their own way contributed to my early odyssey with bamboo.

To the 24 contributing bamboo artists, thank you for sharing your rich designs; I trust your work will continue to flourish. Thank you Bill Alexander, for bringing to light Frederick Law Olmsted's extensive use of bamboo more than a century ago. Special thanks as well to Roger Bakeman, Carolyn Brooks, Ada Lea Birney, Gail and Brian McCarthy, and Mary and Robert Thompson for opening their homes and grounds to our cameras.

For assembling the vision into beautiful three-dimensional form, a spirited thank-you to Dana Irwin, talented art director and long-time lover of bamboo. Gratitude as well to Sandra Stambaugh for her gorgeous photography, and to Olivier Rollin for his outstanding illustrations. From the book's first printing in 2001 through this updated and revised edition, a special thanks to my dear friends Elizabeth Polchow Livingston and Michael JamesLong, both of whom encouraged and supported my far-flung adventures along the bamboo trail.

CAROL STANGLER

bamboo art, craft lectures, and demonstrations

HARRY ABEL

Bamboo water features, fences, and structures

JOE BABB

hand forged ironwork

BAMBOO-SMITHS (Reed Hamilton, Doug and Martha Lingen)

custom gates, fences, and garden structures

CAROLYN BROOKS

creative self-discovery workshops

ELLISTON DESIGN STUDIO (Michael Elliston)

furniture and design

BAMBOO FENCER, INC. (Dave Flanagan)

bamboo fences, gates, screens, and railings

PETER GALLAGHER

bamboo vessels, plates, and sculptural arbors

CAL HASHIMOTO

Bamboo sculpture, private and corporate commissions

ANITA MATOS

horticultural workshops

FRANCISCO PLAZA

fine bamboo furniture and furnishings

OZARK BAMBOO GARDEN (Gordon Powell)

bamboo screens and lamps

JOHN SCHEER

bamboo structures, furniture, decor, and inventions

BARBARA SCHINDLER

bamboo harmony hangers and decor

KAT SEMRAU

contemporary art

ROBERT SMALL

custom bamboo gates, fences, screens

SMITH & FONG CO.

bamboo strand flooring, plyboo, plyboo paneling, veneer

MICHEL SPAAN

custom bamboo arbors, screens, and water features

BROTHER KONOMO UTSUMI

walks for world peace and nuclear disarmament

YUCATAN BAMBOO

"iron" bamboo fences, screens, furniture

—*Only the bamboo*
knows the pure breeze—

Inscription from an ink painting by Zen Master Nanrei Kabori-roshi, Ryoko Temple, Kyoto, Japan, 1974.

Index

Photo Credits

p. 1, 11 (top center right), 25 (bottom), 26, 27 (top left, top right, center left, and center right), 28, 47, 64, 65, 104, 106, 157: Evan Bracken, photographer

p. 3 (bottom right), 5 (second and third from top), 6, 7, 8 (top right, bottom left, and bottom right), 9 (bottom), 10 (bottom), 11 (lower right), 13 (lower left), 14, 24, 27 (lower right), 29 (bottom), 30 (top and lower left), 38 (top and lower right), 40 (bottom center), 42 (top), 50 (top), 53 (bottom), 69 (bottom), 100, 101 (center), 103, 112, 113, 118 (left), 120, 121, 122, 123, 133 (far right), 135 (top right), 143, 158: Carol Stangler, photographer

p. 5 (top): photo by Adam and Sue Turtle, ©Temperate Bamboo Quarterly, used with special permission
p. 8 (top left and center right), 9 (top and center), 10 (top), 11 (bottom left), 20, 157: Michael JamesLong
p. 11 (center right), 69 (top): Gordon Powell
p. 12, 13 (bottom right): Marler Spence
p. 16 (bottom left): Robin McBride Scott
p. 21: Ian Connor
p. 22: Jennifer James-Long
p. 23: Bill Alexander
p. 37, 40 (top): The Bamboo-Smiths
p. 37 (lower right), 114: Dave Flanagan
p. 50 (middle and bottom): Denise Laffan
p. 87: Peter Gallagher
p. 133 (lower left): Robert Small

Usually, you can find the supplies you need for making the projects in Lark books at your local craft supply store, discount mart, home improvement center, or retail shop relevant to the topic of the book. Occasionally, however, you may need to buy materials or tools from specialty suppliers. In order to provide you with the most up-to-date information, we have created a listing of suppliers on our website, which we update on a regular basis. Visit us at www.larkbooks.com, click on "Sources," and then search for the relevant materials. You can also search by book title, vendor, and author name. Additionally, you can search for supply sources located in or near your town by entering your zip code. You will find numerous companies listed, with the web address and/or mailing address and phone number.